# Digital Transformation in Engineering Management

emerald PUBLISHING
ice Publishing

# Digital Transformation in Engineering Management

## Case studies of digital transformation in Africa

Edited by
**Arnesh Telukdarie**

**Published by Emerald Publishing Limited**, Floor 5, Northspring, 21–23 Wellington Street, Leeds LS1 4DL.

ICE Publishing is an imprint of Emerald Publishing Limited

**Other ICE Publishing titles:**
*Digital Twins for Smart Cities: Conceptualisation, challenges and practices*
Li Wan, Timea Nochta, Junqing Tang and Jennifer Schooling.
ISBN 978-0-7277-6600-7
*Energy and Mobility in Smart Cities*
William J Nuttall, David V Gibson, Dariusz Trzmielak and Alejandro Ibarra-Yunez. ISBN 978-0-7277-6425-6
*BIM for Project Managers: Digital construction management*
Peter Barnes. ISBN 978-0-7277-6529-1

A catalogue record for this book is available from the British Library

ISBN 978-1-83608-365-8



Cover photo: Jamesteohart/Shutterstock.com

Commissioning Editor: Michael Fenton
Content Development Editor: Cathy Sellars
Production Editor: Emma Sudderick

Typeset by KnowledgeWorks Global Ltd.
Index created by Lyn Nesbitt-Smith

# Contents

# Praise for *Digital Transformation in Engineering Management*

*Digital Transformation in Engineering Management* is an exceptional work that delves deep into the fascinating world of digitalisation across various industries. The book covers a wide range of industries, including the health care sector, steel, small and medium enterprises (SMEs), mining, data collection, water architecture, and cyber-physical system (CPS). It provides an insightful analysis of each industry's unique characteristics and discusses the complexities of digital transformation at length, offering practical insights into how to navigate the constantly evolving landscape. This book is a must-read for practitioners and researchers who seek to understand the transformative potential of digitalisation. It examines real-world cases that provide valuable insights into the impact of digital innovation. Overall, *Digital Transformation in Engineering Management* is an exceptional contribution to the field of digital transformation in the twenty-first century.

**Hlobisile Mahure**, software developer and researcher at the University of Johannesburg, South Africa.

# Dedication

This book is a tribute to the outstanding individuals who are at the forefront of driving change through innovation and to the professionals who are tirelessly working towards harnessing the immense power of digitalisation. Their hard work and dedication are instrumental in shaping a future where technology plays a pivotal role in enhancing our lives in ways we have yet to imagine. We are immensely grateful to these exceptional individuals for their relentless efforts to push the boundaries of what's possible and for their unwavering commitment to creating a better, more advanced future. We hope that their work will continue to inspire and contribute to the advancement of technology in the years to come.

# Acknowledgments

We extend our heartfelt appreciation to the University of Johannesburg's Process, Energy, and Environmental Technology Station (UJ PEETS) for their valuable contributions to this book. UJ PEETS has played a pivotal role in enhancing the content and insights, bringing specialised knowledge and expertise to enrich the discourse on our subject matter. We are grateful for their unwavering support in our endeavour.

We also wish to express our sincere gratitude to the Manufacturing, Engineering, and Related Services Sector Education and Training Authority (merSETA) for their significant contribution to this book. merSETA's expertise and support have greatly enriched the content and understanding of the subject matter. We acknowledge and appreciate their valuable partnership in bringing this project to fruition.

**Chapter 1:** The authors extend their heartfelt thanks to the team at CHIETA for their valuable cooperation and insights that helped explore the implementation of business intelligence in the chemical and manufacturing industries.

**Chapter 2:** The authors express their sincere appreciation to Lubuto Hospital and the dedicated health care professionals in Gauteng for generously sharing with the study their experiences of the digital transformation of health care maintenance management.

**Chapter 3:** BBT Steel Manufacturers are acknowledged for their valuable input and contributions to the analysis of digital transformation in the steel industry.

**Chapter 4:** The authors extend special thanks to the FoodBev SETA and participating SMEs in South Africa for their candid and insightful contributions that enabled the exploration of the challenges and prospects of SMEs in digital transformation.

**Chapter 5:** The authors would like to express their sincere gratitude to the organisation that participated in gathering childhood development data. These organisations provided valuable insights into the role of digitalisation in this crucial domain. The authors would also like to extend their thanks to

Mr. Regionald Mongwe, the sole developer of the CWTT application.

**Chapter 7:** The experts in water architecture are acknowledged for their valuable insights that enrich the understanding of the digital transformation in water-related industries.

**Chapter 8:** The authors extend their special thanks to the multinational corporations that participated in the discussions around reproducible and sustainable cyber-physical system implementation. Their contributions were critical to the deeper exploration of this topic.

**Chapter 9:** Finally, the FoodBev architecture experts in South Africa are acknowledged for their insights into digital systems analysis that enhanced the understanding of digital transformation in this sector.

# Preface

*Digital Transformation in Engineering Management* is an informative and thought-provoking book that takes readers on a fascinating journey through various industries. The book covers a wide range of industries, including manufacturing, health care, steel, food and beverage, childhood development, water architecture and multinational corporations. In each chapter, the authors delve deep into the heart of these industries, offering unique insights into the innovative ways in which digitalisation is transforming companies, enhancing their efficiency, and sustainability. The book is a testament to the collaborative efforts of many organisations and individuals who generously shared their knowledge and experiences.

As we read through the narratives woven within these pages, we are inspired to embrace the future that technology beckons. The exploration of these industries showcases the immense benefits of technology in enhancing their operations and ensuring that they remain competitive in an ever-evolving world. *Digital Transformation in Engineering Management* is a must-read for practitioners, researchers, and anyone with a curious mind. It highlights the importance of embracing digitalisation and the benefits it can bring to businesses, individuals and society. Overall, this book will leave readers with a deeper understanding of the impact of technology on various industries and inspire them to recognise the transformative potential of digitalisation.

# About the editor

Professor Arnesh Telukdarie is a well-known expert in digital businesses and a Professor at the Johannesburg Business School, University of Johannesburg. He is an Adjunct Professor at Johns Hopkins University and a Visiting Scholar at London South Bank University. Professor Telukdarie is also the research chairperson at the Johannesburg Business School, where he focuses on Industry 4.0, Society 5.0, sustainable development, business optimisation and manufacturing systems. He has provided consultation services to Transnet, Johannesburg Water, and the Council for Scientific and Industrial Research (CSIR), among others.

# About the contributors

**Professor Leila Patel** is a renowned researcher and former Director General of Social Welfare in South Africa's Mandela government. With a B2 rating, she is a globally recognised expert in social development and has published extensively on topics such as social welfare policy, social protection and socio-economic rights. She has worked in academia, government, nonprofit organisations and private sector social involvement initiatives. Among her recent publications are *Development, Social Policy and Community Action: Lessons from Below* (2017), *Social Welfare and Social Development* (2015), and *Social Protection in Southern Africa* (2014). Professor Patel's exceptional contributions to the field have earned her numerous awards, including the Distinguished Woman Scientist Award in the Humanities and the Social Sciences (2014), the Distinguished Alumni Award from the University of the Western Cape (2015) and the Science for Society Award from the Academy of Science of South Africa (2020).

**Professor Lauren Graham** is a development sociologist who is deeply committed to social justice. As an academic, her work revolves around enhancing opportunities for individuals to realise their full potential. Lauren's research has primarily focused on marginalised and excluded groups, including unemployed youth, individuals with disabilities, and children growing up in adverse circumstances. She holds the position of Professor at the University of Johannesburg and has been designated an NRF-rated scholar. Lauren is also the Director at the Centre for Social Development in Africa and serves as the Chairperson for the International Consortium for Social Development: Africa Branch.

**Dr. Michael Sishi** is a highly distinguished professional with a Ph.D. in IoT and AI from the University of Johannesburg. He has played a significant role in the company's digital transformation, with a particular emphasis on Manufacturing Execution Systems (MES) for IoT. Dr. Sishi is also a Fellow of the University of Johannesburg Business School, South Africa. He has contributed extensively to various technology journal articles and book chapters.

**Dr. Megashnee Munsamy** holds a Ph.D. and a master's degree in chemical engineering (with cum-laude). She currently serves as a Lecturer at the Johannesburg Business School, University of Johannesburg.

**Thabile Dube**, with a master's degree in Tourism and Hospitality Management from the University of Johannesburg, currently holds the position of Project Manager and Research Coordinator at the Johannesburg Business School.

**Dr. Eyad Buhulaiga** holds a Doctor of Philosophy (Ph.D.) in Engineering from the University of Johannesburg. He is an alumnus of Nova Southeastern University, where he obtained a master of information technology, and King Fahd University of Petroleum and Minerals where he obtained a bachelor's in information and computer sciences. Eyad has attended several leadership programs and earned professional certifications from leading universities including MIT, Stanford University, and London Business School.

**Joseph Mwanza** earned his bachelor of engineering degree in industrial engineering from the National University of Science and Technology in Zimbabwe in 2001. He later completed his master of engineering degree in manufacturing engineering and operations management at the same institution in 2013. Currently, he is a doctoral candidate in engineering management at the University of Johannesburg, focusing his research on leveraging digital technologies to develop cost-effective optimisation solutions for business systems, with a specific emphasis on enhancing health care systems.

**Dr. Bupe Gertrude Mwanza** is a Postdoctoral Research Fellow at the Johannesburg Business School, University of Johannesburg. With over a decade of experience in higher education institutions, her expertise lies in teaching, research, and quality assurance. She earned her Ph.D. in engineering management from the University of Johannesburg and has contributed research to various peer-reviewed journals, book chapters, and other scholarly publications.

**Logistic Makoni** is a skilled and dedicated individual who currently serves as a temporary Tourism Lecturer at the School of Tourism and Hospitality (STH), University of Johannesburg. In addition, he is a temporary Research Coordinator at the Johannesburg Business School. As a young and enthusiastic researcher and academic, Logistic is deeply committed to addressing developmental challenges in Africa, especially in the areas of planning and policy-related issues. Logistic is in the final year of his Ph.D. program at the University of Johannesburg, and his strong motivation drives him to make a positive impact in his field of expertise.

**Kedibone Moroane** holds a master's degree in business administration from the renowned Witwatersrand University in South Africa. Her skills range from policy analysis, human resource management, research, and planning. She contributed as a policy analyst at the Chamber of Mines of South Africa and thereafter held the position of Research and Skills Planning Manager at the Chemical Industries Education and Training Authority (CHIETA). Currently, she holds the position of Executive Manager in the field of research and planning at CHIETA.

**Reginald Mongwe** serves as a research assistant at the University of Johannesburg (Johannesburg Business School), specialising in software development. He is currently pursuing a master's degree in engineering at the Durban University of Technology.

**Xolani Maphisa** is a dedicated data science enthusiast who holds a master's degree in computational physics from the University of Johannesburg. He is currently engaged in a Ph.D. program focused on Digital Transformation at the Johannesburg Business School.

**Hlobisile Mahure** is a Software Developer and Researcher at the University of Johannesburg, South Africa. Her core areas of expertise include software development, human-computer interaction, web programming, machine learning, and research methods.

# List of figures and tables

## Figures

## Tables

Arnesh Telukdarie
ISBN 978-1-83608-365-8
https://doi.org/10.1108/978-1-83608-362-720241001


Chapter 1

# An introduction to digitalisation in the field of engineering management

A Telukdarie and H Mahure

## Introduction

The field of engineering management, coupled with digital transformation, is an intriguing yet complex paradigm. It encompasses a range of disciplines, including project management, process optimisation, innovation management and systems engineering. These areas work together to improve organisational performance and competitiveness in this increasingly digital age. From implementing cutting-edge technologies to creating innovative business models, engineering management and digital transformation require a deep understanding of both the technical and managerial aspects of modern organisations. However, achieving an integrated, digitally engineered business is challenging. The transition to digital solutions requires practical guidance for young professionals to build confidence as they embark on their digital journey. To bridge this gap, this book offers diverse and detailed perspectives through real-world case studies, documenting the journey of industry professionals in their digital transformation journey. The case studies detail the path of effective digital transformation, showcasing the challenges and successful strategies employed. The book aims to share practical experiences and learnings from these case studies to assist in future digital implementations. It offers readers a unique opportunity to learn from the real-world experiences of industry professionals, providing them with the insights required to make informed decisions and drive successful digital transformation in their organisations.

Digital transformation is a process that involves the transformation of business operations and activities using integrated digital technologies. The process is carried out by taking a systems-based approach that can address complex challenges by incorporating multiple dimensions. Engineering management plays a crucial role in digital transformation by providing a framework for managing the design, development and implementation of integrated digital solutions. Through engineering management, businesses can effectively integrate digital technologies into their operations, optimise their processes and achieve their goals more efficiently. The goal of digital transformation is to create a more agile, efficient and innovative organisation that can stay ahead of the competition in an increasingly digital world. The architecture of such a system involves various functions that are integrated into a cohesive whole, as illustrated in Figure 1.1.

Figure 1.1 Systems-based approach: engineering management (author's own)

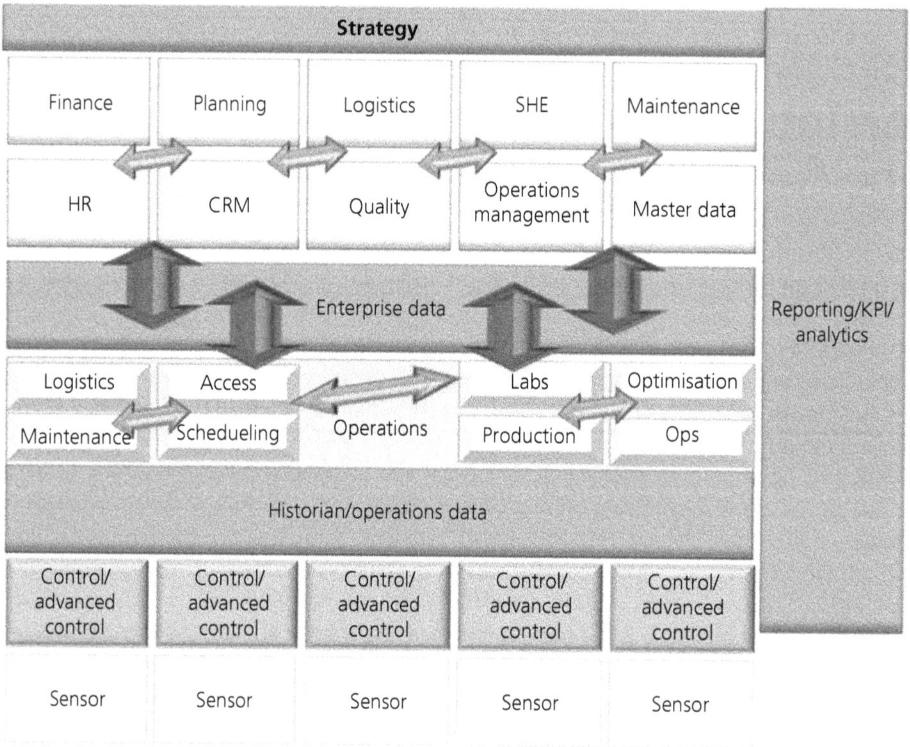


The advancement of the fourth industrial revolution (4IR) relies on several fundamental aspects, including horizontal, vertical and total business integration (Serumaga-Zake *et al.*, 2021; Peerally *et al.*, 2022). The collection of case studies herein illustrates these core principles in action. In addition, it is essential to integrate all business functions, such as operations, maintenance, planning, finance, human resources, quality, safety and research, into a cohesive and synergistic system to achieve optimum business optimisation. The case studies cover various aspects of the business world, including social system development, business intelligence and specialist toolset development, in both the public and private sectors. This book focuses on system thinking and consulting, as illustrated in Figure 1.2. It aims to develop system models that break down barriers to individualistic modelling and optimisation, advancing multisystem integration and optimisation.

In the age of 4IR, system thinking can deliver significant value to a single business, a value chain or an integrated system. The objective of this collection of case studies is to lead the readers towards an system paradigm, grounded in digital technology to advance future-fit thought processes. This book includes chapters that cover an extensive range of topics, including manufacturing, health care, steel, food and beverage, childhood development, water

Figure 1.2 System approach (author's own)

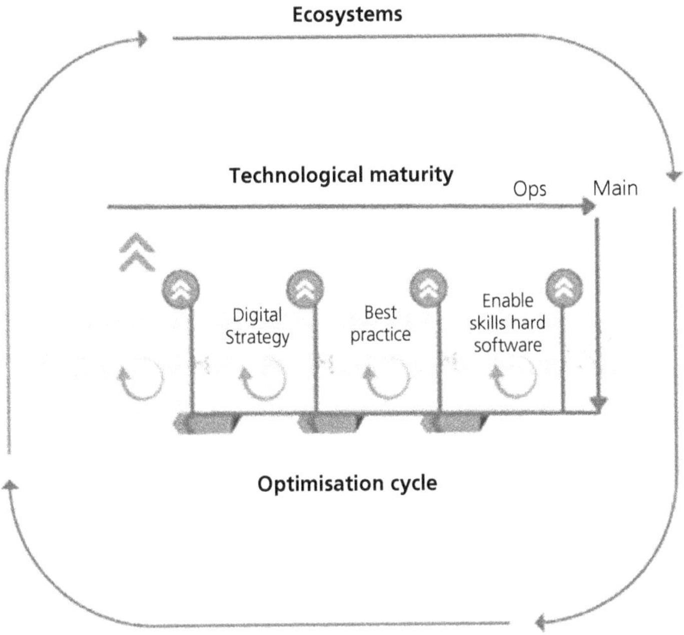


architecture and multinational corporations. Each chapter is dedicated to exploring the integration of digitalisation in different industries, providing unique insights into the innovative ways in which digitalisation can transform businesses, and enhance efficiency and sustainability. Each chapter further details how digitalisation is leveraged to solve industry challenges such as operational inefficiencies. The book provides an in-depth understanding of how leveraging digitalisation and innovation can improve efficiency and yield results throughout. The following is a breakdown of each chapter's uniqueness as well as the structure of the book.

The book comprises nine chapters. The first chapter provides a detailed discussion of the implementation of a business intelligence (BI) system at the Chemical Industries Education and Training Authority (CHIETA) in South Africa. The chapter offers a comprehensive overview of the role of BI in enhancing skills in forecasting, data analytics and decision-making processes. It also highlights the significance of data-driven strategies for aligning training programs with labour market requirements and achieving long-term success in skills development initiatives.

Moving forward, Chapter 2 of the book delves into the topic of digital transformation in the health care sector. The chapter presents a case study of Lubuto Hospital and how it dealt with the challenges that arose during the COVID-19 pandemic. The chapter elaborates on how integrated digital innovation is able to achieve cost-effective optimisation, outlining the steps taken to develop the digital solution. It further demonstrates how innovative methods and digital tools can be used to solve pressing business challenges systematically. This case study aims to help readers in the industry develop their problem-solving skills and gain insights into the importance of digital transformation in the health care sector.

In the third chapter of the book, the focus is on Steel Manufacturers, and the challenges faced in forecasting due to a gap between their manufacturing and business systems. The chapter highlights how integrated digitisation played a crucial role in bridging this gap by utilising existing investments to create smart enterprises capable of efficiently managing current and future steel. The study adopts Industry 4.0 integrated technologies and proposes a framework that can be implemented in the steel sector to achieve these objectives.

Chapter 4 sheds light on the challenges and possibilities faced by small and medium-sized enterprises (SMEs) in South Africa's food and beverage manufacturing industry during their shift towards digitalisation. The chapter emphasises how SMEs in this industry face various challenges, including competition with larger businesses, financial limitations, limited access to technology, credit difficulties, skill shortages, policy restrictions and infrastructure inadequacies. However, the 4IR offers a solution for improved manufacturing, financial management and environmental sustainability through digitalisation. The chapter highlights the importance of digital transformation for SMEs to ensure their sustainability and competitiveness in South Africa's ever-changing food and beverage sector.

Chapter 5 of the book delves into the subject of digitalisation and its role in gathering data in the context of childhood development in Africa. The chapter provides an insightful account of the journey from the traditional paper-based data collection system to the development of a sophisticated digital tool for data collection and analysis. The case study highlights the challenges and opportunities involved in the transition to a digital system, including the advantages of greater efficiency, accuracy and speed of data processing.

In Chapter 6, the book examines the challenges faced by Company X in the process of digital transformation, emphasising the importance of a well-planned and integrated digital strategy. The chapter describes the various limitations of standalone digital solutions and the importance of alignment with business objectives. The study documents the development of a comprehensive digital business strategy aimed at achieving total business visibility, automation of non-value-adding activities and digitalisation of various aspects. The case emphasises the importance of effective change management for successful digital transformation, including the need for clear communication, strong leadership and employee engagement. This study provides valuable insights into the potential benefits of digitalisation and highlights the need for an integrated approach to digital transformation.

Chapter 7 of the book delves into the topic of digitalisation in water architecture and explores how advanced 4IR technologies are utilised to address the pressing issues of water scarcity and distribution in the city of Johannesburg. The case study provides a detailed account of the challenges faced by the city's water management authorities and how they leveraged digital solutions to overcome these hurdles.

Moving on to Chapter 8, the book delves into the implementation of a cyber-physical system (CPS) in a multinational enterprise (MNE) energy company. The chapter provides an in-depth analysis of the challenges faced by the company in the implementation of the CPS and how they are overcome using a range of advanced technologies.

Chapter 9 of the FoodBev architecture analysis in South Africa explores the use of 4IR technologies to enhance the efficiency, effectiveness, and impact of the FoodBev SETA. The case study highlights the development of integrated system analysis to streamline processes and elevate overall performance, as FoodBev SETA aims to remain competitive and relevant in a rapidly evolving technological landscape.

Overall, this book is an excellent resource for anyone interested in learning about the practical implementation of an integrated digital business in various industries. It offers valuable insights into how digitalisation can enhance business efficiency and sustainability. The book provides a diverse and in-depth perspective on digital transformation through real-world case studies of industry professionals who navigate the complexities of digitalisation. These case studies give valuable insights into the effective path of integrated digital transformation, highlighting the challenges, pitfalls and successes of real industry professionals. The book aims to assist future generations of professionals in their digitalisation journey by sharing the lessons learned from practical experience.

Essentially, the book provides a roadmap for effectively implementing digital transformation in the engineering sector, with diverse examples by way of case studies. It showcases real-world examples and practical guidance, making it a valuable resource for professionals seeking to navigate the complexities of digital integration with greater confidence and success. Readers will benefit from understanding the different strategies required to successfully implement business systems, and engineers will gain insights into the design process specific to each area of interest. The book also contributes to building Africa's business environment and development through practical examples of local solutions to a range of business challenges. Exploring Digitalisation in Engineering Management is an essential guide for both practitioners and researchers who want to understand the transformative potential of digitalisation.

## REFERENCES

Serumaga-Zake JM and van der Poll JA (2021) Addressing the impact of fourth industrial revolution on South African manufacturing small and medium enterprises (SMEs). *Sustainability* **13(21)**: 11703.

Peerally JA, Santiago F, De Fuentes C and Moghavvemi S (2022) Towards a firm-level technological capability framework to endorse and actualise the Fourth Industrial Revolution in developing countries. *Research Policy* **51(10)**: 104563.

Arnesh Telukdarie
ISBN 978-1-83608-365-8
https://doi.org/10.1108/978-1-83608-362-720241002


Chapter 2

# Business intelligence implementation at CHIETA: addressing skills demands in the chemical and manufacturing industries

A Telukdarie, K Moroane and X Maphisa

## Introduction

Lizzy Tefo is a highly motivated and passionate individual who has dedicated her career to the field of skills development and training. As the Director of Skills Planning and Development at the Chemical Industries Education and Training Authority (CHIETA), Lizzy plays a crucial role in shaping the future of the chemical and manufacturing industries in South Africa. With a deep understanding of the challenges faced by the sector, Lizzy is committed to facilitating skills development and addressing the skills needs within the industry. She recognises that the success of CHIETA's initiatives depends on their ability to accurately identify and respond to the evolving skills landscape.

Lizzy is the driving force behind CHIETA's strategic planning efforts, with a particular focus on the Sector Skills Plan (SSP). She understands the significance of the SSP as a roadmap for guiding CHIETA's skills development interventions. To ensure the SSP's effectiveness, Lizzy leads the annual analysis of Workplace Skills Plans (WSPs) and Annual Training Reports (ATRs), meticulously examining the data and insights they provide. Through her dedication and expertise, Lizzy leverages the findings from the WSPs and ATRs to gain a comprehensive understanding of the sector's current and future learning and qualification needs. She recognises that the infrequent assessment of skills within the sector poses a significant challenge, given the rapidly changing demands of the labour market. Lizzy is determined to overcome this challenge and make the SSP a more dynamic and proactive tool for skills forecasting.

Lizzy actively collaborates with industry stakeholders (University of Johannesburg Business School), conducts research, and explores innovative approaches to enhance CHIETA's skill forecasting capabilities. She understands the importance of leveraging data analytics and industry insights to predict future skills needs accurately. By doing so, Lizzy aims to ensure that CHIETA and the entire SETA network can adapt their skills training programmes to meet the evolving demands of the labour market effectively. Lizzy's passion, expertise, and strategic thinking will be instrumental in navigating the challenges faced by CHIETA.

Her role as the Director of Skills Planning and Development empowers her to drive change, shape industry collaborations, and influence the allocation of funds towards priority skills development initiatives. Lizzy's journey will showcase the critical role of proactive skills planning in equipping the chemical and manufacturing industries with the necessary competencies for sustainable growth and competitiveness.

## The Chemical Industries Education and Training Authority (CHIETA)

The Chemical Industries Education and Training Authority (CHIETA), established by the Skills Development Act in 1998, plays a vital role in facilitating skills development and addressing the skills needs within the chemical and manufacturing industries in South Africa. As one of the 21 Sector Education and Training Authorities (SETAs) in the country, CHIETA is responsible for implementing various learning programmes, including learnerships, skills programmes, internships and other strategic initiatives to enhance the skills of the workforce.

To support the development of skills in the sector, CHIETA allocates a significant portion of the Skills Development Levies (SDL) obtained from the chemical industry. Approximately 70% of the SDL is distributed back to member companies through mandatory and discretionary grants. Mandatory grants are provided to companies that fulfil the requirements, ensuring that they receive the essential support for skills development. These grants serve as a critical source of funding for companies to enhance the skills of their employees and meet industry standards. In addition to the mandatory grants, CHIETA also offers discretionary grants. These grants are awarded at the discretion of CHIETA for skills development projects linked to scarce and critical skills. By providing discretionary grants, CHIETA supports initiatives that address the specific skills needs of the sector, fostering the growth and competitiveness of the chemical and manufacturing industries.

This flexible funding mechanism allows CHIETA to adapt to the evolving skills landscape and prioritise investments in areas that have the greatest impact on the industry. As a trusted partner in skills development and training for the chemical sector, CHIETA funds various occupational programmes as well as certain Technical and Vocational Education and Training (TVET) sector and higher education programmes. The allocation of funds is guided by CHIETA's well-researched SSP, which documents the hard-to-fill skills and hard-to-fill positions within the sector. The SSP provides a roadmap for CHIETA's skills development interventions, ensuring that resources are directed towards areas of critical need.

However, the infrequent assessment of the SSP within the sector poses a significant challenge for CHIETA. The labour market is constantly evolving, driven by technological advancements, industry shifts and changing economic conditions. Without regular evaluations, there is a risk of the skills gaps widening and the workforce being ill-prepared to meet industry demands. To overcome this challenge, CHIETA recognises the importance of adopting a proactive approach to predicting future skills needs. By leveraging data analytics, conducting comprehensive research and engaging with industry stakeholders, CHIETA aims to enhance its skills forecasting capabilities to stay ahead of emerging skills requirements.

## Enhancing CHIETA's planning for skills development

Figure 2.1 (see Exhibits) provides a visual representation of the information process cycle that informs CHIETA's planning for skills development. The cycle outlines the key steps involved in gathering, analysing, and utilising information to guide strategic decision-making. The colour coding of the arrow indicates whether CHIETA performs that specific process well (green), can do better (yellow), or is not doing well (red). By improving all the cycles in this process, CHIETA can ensure that its planning efforts are well-informed, aligned with sectoral and national priorities, and effectively address the industry's skills needs.

Research and Strategic Planning: The cycle begins with research and strategic planning, where CHIETA establishes objectives, targets and outcomes based on national priorities, the Department of Higher Education and Training's (DHET) Service Level Agreement (SLA) and sectoral research reports (tracer studies, impact reports, annual reports and SSPs). This step ensures that CHIETA's planning is grounded in a thorough understanding of the industry's current and future requirements.

Indicator Development: To effectively monitor and evaluate the impact of its programmes, CHIETA develops indicators that serve as benchmarks for measuring progress and success. These indicators are derived from the programmes implemented by CHIETA and are influenced by the SLA with DHET. The development of indicators ensures that CHIETA's performance can be tracked and assessed against predetermined targets.

Data Collection: Data collection is a critical process in the information cycle. CHIETA gathers information on applicants and programme awardees, including learnerships, apprenticeships, bursaries, skills programmes, work-integrated learning (WIL), lecturer development, and support for small, medium and microenterprises (SMMEs). This data is collected at the unit record level and serves as the foundation for strategic priorities and decision-making. The collected data forms the basis for the development of strategic priorities. The data also consist of the WSP and ATRs that are used in the development of the SSP and inform priority funding based on the identified occupational needs and training. The data collected also includes information on discretionary grant funding. CHIETA aligns its funding decisions with industry needs identified in the Sector Skills Plan (SSP).

Dedicated impact assessment: Dedicated impact assessment plays a crucial role in CHIETA's planning process, allowing for a comprehensive evaluation of various projects and programmes against predefined indicators. These assessments aim to determine the return on investment for the different funding vehicles employed by CHIETA to support skills development initiatives. However, currently, CHIETA does not have dedicated impact assessment surveys integrated into this process. Instead, the organisation conducts ad hoc impact research to assess the effectiveness of its interventions. This ad hoc approach lacks the systematic and continuous evaluation needed to fully gauge the impact of their programmes. By embracing a more regular and comprehensive approach to impact assessment, CHIETA can strengthen its ability to drive meaningful data-driven decision-making in the pursuit of effective skills development.

Figure 2.1 emphasises the flow of information and highlights the importance of data collection for effective monitoring and evaluation of skills interventions. To enhance the information flow and strengthen CHIETA's planning efforts, Lizzy/CHIETA recognised the need for investments in data collection, the development of processes for data analysis and aggregation of impact surveys, and the establishment of a clear plan of action. These steps will ensure that CHIETA's skills planning is comprehensive, evidence-based, and aligned with sectoral and national priorities.

## CHIETA and University of Johannesburg collaboration

Recognising the importance of robust data collection and analysis for effective skills planning, CHIETA proactively established a strategic partnership with a renowned research group at the University of Johannesburg's Johannesburg Business School (UJ-JBS). This collaborative alliance aimed to leverage the expertise of both organisations to enhance CHIETA's skills planning initiatives. Led by Lizzy Tefo, the Head of Skills Planning and Development at CHIETA, the collaboration between CHIETA and the UJ-JBS research group focused on developing a cutting-edge business intelligence tool specifically designed to support CHIETA's skills planning efforts. Lizzy, with her profound understanding of the industry's skills landscape and CHIETA's strategic objectives, played a pivotal role in driving this joint venture.

The partnership commenced with a comprehensive analysis of CHIETA's existing data collection methods, systems, and processes. The UJ-JBS research group conducted an in-depth evaluation to identify areas for improvement and refinement. This assessment facilitated the identification of critical data gaps, data quality issues, and opportunities for leveraging advanced analytics and modelling techniques. Building upon these insights, the UJ-JBS research group collaborated closely with Lizzy and her team to design and develop a sophisticated business intelligence tool tailored to CHIETA's unique requirements.

## Business intelligence

Business intelligence (BI) refers to the comprehensive utilisation of organisational resources, including data, applications, people, and processes, to enhance knowledge, implement strategies, and adapt to dynamic environments (Bhosale *et al.*, 2021). BI solutions, as described encompass techniques and practices designed to analyse data generated from an organisation's day-to-day operations. These solutions enable organisations to identify trends and patterns, create reports, dashboards, and perform predictive analytics, thus gaining a competitive advantage in their industry (Niu *et al.*, 2021). Recent literature highlights the significance of effective data utilisation for organisations to drive innovation, identify new opportunities, and gain a competitive edge in the market (Sivarajah *et al.*, 2017). As a strategic framework, BI is increasingly crucial in supporting business strategies (Alnoukari and Hanano, 2017). It is considered one of the most critical technologies that enable organisations to transform vast amounts of opaque data into valuable information, empowering them with substantial capabilities. BI, as an analytical tool, converts internal and external data into relevant knowledge that supports decision-making processes (Wieder and Ossimitz, 2015).

## Business intelligence pipeline

The UJ-JBS research group followed a systematic approach shown in Figure 2.2 (see Exhibits) to build the BI tool. The initial step was to explore and understand the collected data to determine the necessary preparations for its intended use. Data preprocessing played a critical role in addressing common anomalies and ensuring data accuracy. These anomalies included inadequate data profiling, missing or incomplete data, invalid data values, inconsistent naming and addressing conventions, and inconsistencies across different systems or tables. To carry out the preprocessing stage, the UJ-JBS research group utilised their expertise in data science, leveraging the capabilities of the Python programming language. They specifically employed the Pandas library, widely popular among data scientists, which offers functions for analysing, cleaning, exploring and manipulating data sets. Another Python package, Pandas-Profiling, was used to facilitate exploratory data analysis (EDA) with concise and rapid results. In addition, the Pyodbc library was utilised to access the Microsoft SQL database, where the preprocessed data would be stored.

During the preprocessing stage, the UJ-JBS research group meticulously addressed the identified inconsistencies in the data. They rectified missing column names, corrected misspelled entries, and resolved inconsistent unique identifiers to ensure data consistency. By addressing these issues before loading the data into the database, the research group ensured a solid foundation for further analysis and visualisation. Following preprocessing, the UJ-JBS research group proceeded with data normalisation.

Data normalisation is a process that standardises data by removing redundant or inconsistent information, ensuring consistency across different sources. The research group organised the data based on assigned attributes, adhering to a well-structured data model. Once the data was normalised, the UJ-JBS research group leveraged their expertise in BI tools to integrate the preprocessed and normalised data into Microsoft Power BI. Power BI is a cloud-based analytical tool that enables self-service business intelligence, allowing users to perform data slicing and extraction and transform them into various data visualisations. The research group utilised Power Query Editor, a component of Power BI, to write SQL queries, enabling efficient table joins and exporting relevant data to Power BI for visualisation.

Multiple dashboards were created using Microsoft Power BI (see appendix Figures 2.3–2.7) by the UJ-JBS research group. These dashboards provided comprehensive and intuitive representations of the data, empowering decision-makers at CHIETA with valuable insights for informed decision-making. The expertise of the research group in data analysis and visualisation ensured that the BI tool effectively communicated the analysed information to CHIETA and aided in better skills planning and decision-making processes.

Through their meticulous data preprocessing, normalisation and visualisation efforts, the UJ-JBS research group successfully built a robust and user-friendly business intelligence tool that empowered CHIETA with the essential knowledge and insights to address changes, drive innovation and make informed decisions regarding skills development. The collaborative effort between the UJ-JBS research group and CHIETA exemplified the power of leveraging advanced data analytics and BI tools to support organisational objectives and gain a competitive edge in the industry.

## Conclusion

By harnessing the power of BI, Lizzy and CHIETA were able to analyse data effectively, identify emerging skills needs, forecast future demand and assess the impact of interventions. This enabled

CHIETA to align its skills development initiatives with industry demands and drive innovation. The adoption of BI as a strategic framework empowered CHIETA to convert vast amounts of data into actionable knowledge, enhancing their capabilities and ensuring their long-term success in a dynamic market. The tool integrated multiple data sources, including WSPs, ATRs, labour market data and other relevant industry-specific information. The business intelligence tool incorporated advanced data analytics capabilities, enabling CHIETA to extract meaningful insights and trends from the collected data. It provided dynamic visualisations, customisable dashboards, and interactive reports, empowering CHIETA's decision-makers with real-time information for informed skills planning and interventions. Moreover, the collaboration between CHIETA and the UJ-JBS research group facilitated capacity building within CHIETA's team. Training workshops and knowledge-sharing sessions were conducted to equip CHIETA staff with the necessary skills and knowledge to effectively utilise the business intelligence tool.

The successful implementation of the business intelligence tool revolutionised CHIETA's skills planning processes. It enabled Lizzy and her team to perform comprehensive data analysis, identify emerging skills needs, forecast future demand and assess the impact of interventions with greater accuracy and efficiency. Ultimately, this strategic partnership between CHIETA and the UJ-JBS research group exemplifies the commitment to evidence-based decision-making and continuous improvement in skills planning.

## Exhibits

Figure 2.1 CHIETA's Information planning cycle (author's own)

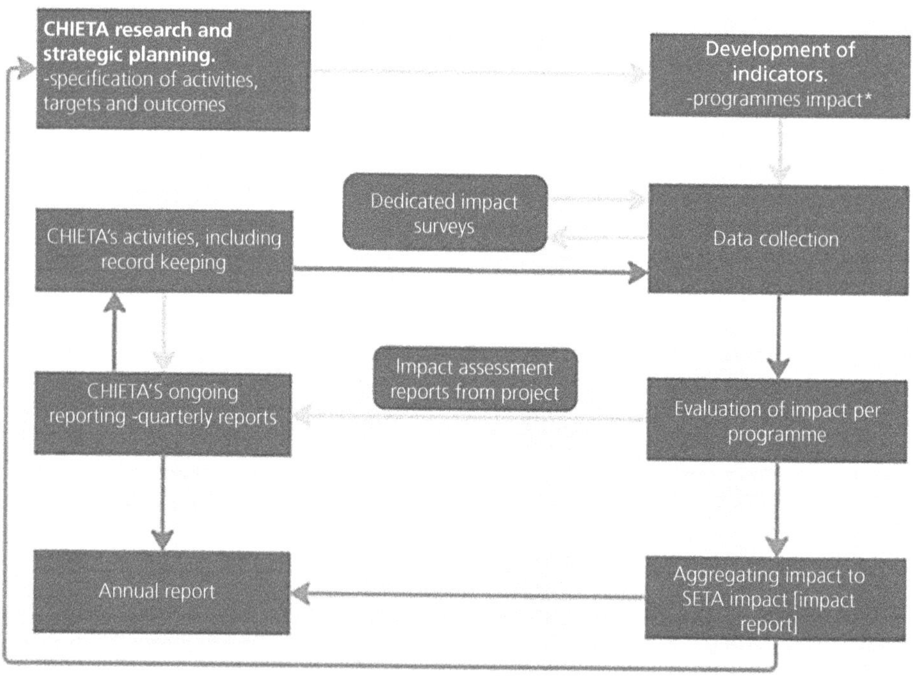

Figure 2.2 Business intelligence process (author's own)

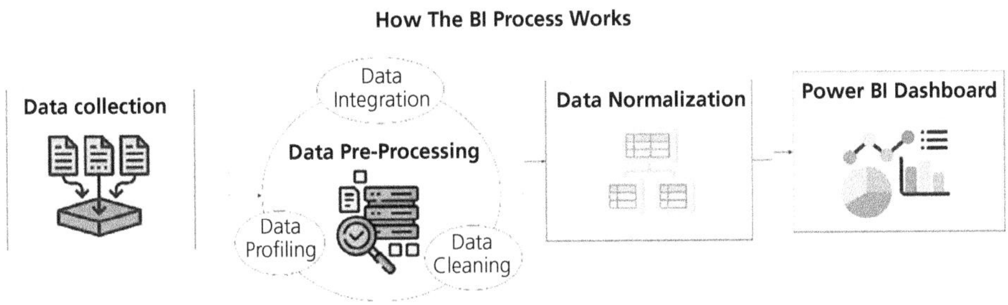


Figure 2.3 Dashboard representing organisations that submitted work skills plan (WSP) to CHIETA (author's own)

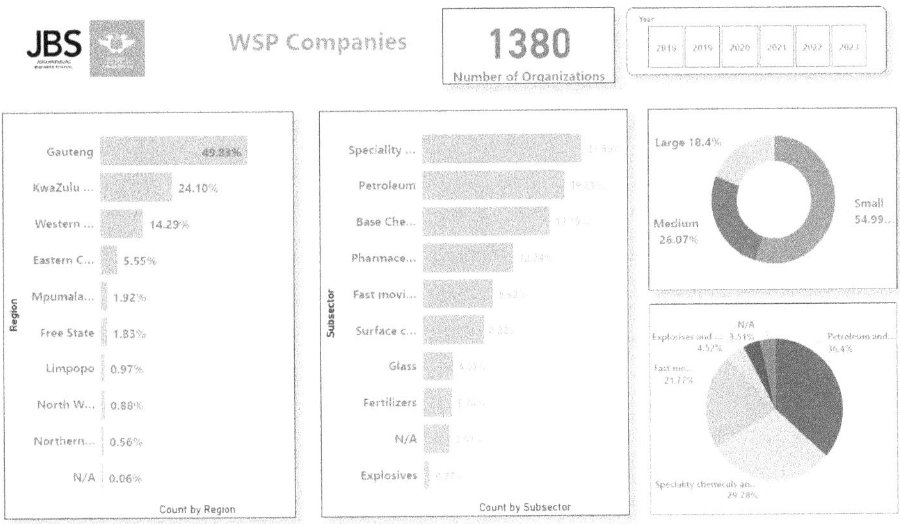

Figure 2.4 Dashboard illustrating CHIETA's levy paying companies (author's own)

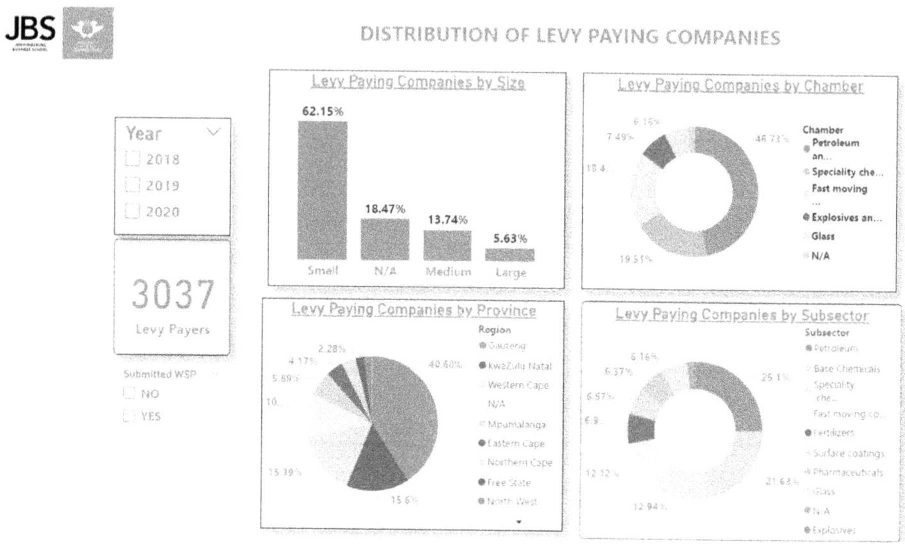


Figure 2.5 Dashboard illustrating levy-paying companies and levies paid (author's own)

JBS

TOTAL LEVIES PAID by REGION/Province

| Year | 2018 | | 2019 | | 2020 | |
|---|---|---|---|---|---|---|
| Region | Levies Paid(%) | Levy Payers(%) | Levies Paid(%) | Levy Payers(%) | Levies Paid(%) | Levy Payers(%) |
| Eastern Cape | 3.54% | 4.08% | 3.54% | 4.08% | 3.53% | 4.34% |
| Free State | 0.60% | 2.21% | 0.60% | 2.21% | 0.37% | 2.42% |
| Gauteng | 69.22% | 40.57% | 69.22% | 40.57% | 68.81% | 40.90% |
| KwaZulu Natal | 11.50% | 15.74% | 11.50% | 15.74% | 11.84% | 15.32% |
| Limpopo | 0.03% | 0.11% | 0.03% | 0.11% | 0.03% | 0.11% |
| Mpumalanga | 0.58% | 5.85% | 0.58% | 5.85% | 0.58% | 5.97% |
| N/A | 1.40% | 10.26% | 1.40% | 10.26% | 1.49% | 10.98% |
| North West | 1.72% | 1.84% | 1.72% | 1.84% | 1.55% | 1.81% |
| Northern Cape | 0.30% | 3.64% | 0.30% | 3.64% | 0.32% | 3.38% |
| Western Cape | 11.12% | 15.70% | 11.12% | 15.70% | 11.48% | 14.78% |

Figure 2.6 Dashboard illustrating employees in companies that fall under CHIETA (author's own)

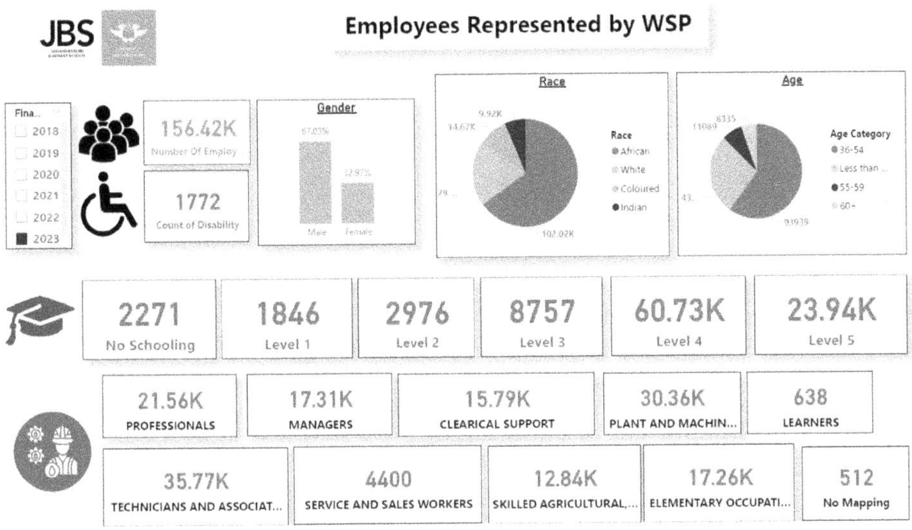


Figure 2.7 Dashboard illustrating skills that were reported as hard to fill (author's own)

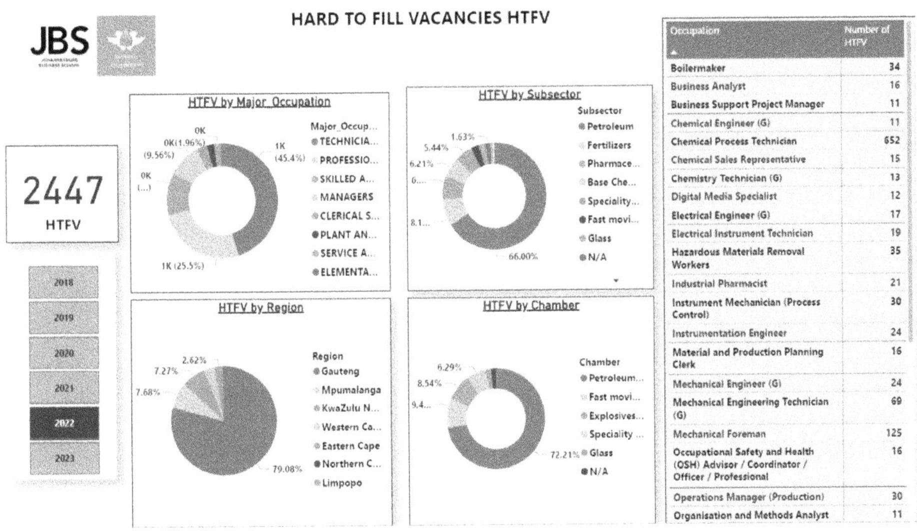

# Teaching notes

## Synopsis

This teaching note guides instructors in facilitating a case study discussion on the implementation of a BI system at the Chemical Industries Education and Training Authority (CHIETA) in South Africa. The case study focuses on the challenges faced by CHIETA in addressing rapidly changing skills demands. It explores the role of BI in enhancing skills forecasting, data analytics, and decision-making processes within the context of skills development and training authorities. Students will gain insights into the importance of data-driven strategies in aligning training programmes with labour market requirements and achieving long-term success in skills development initiatives. Students will also delve into CHIETA's complexities in its mission to develop economic growth skills. They will analyse the limitations of infrequent skills assessments. Through the implementation of a BI system, students will explore how CHIETA can enhance its capabilities in predicting future skills needs. This will improve the efficiency of funding allocation and ultimately align skills training programmes with the rapidly evolving demands of the labour market. The case study prompts students to think critically and develop recommendations for CHIETA to effectively leverage BI and address skills development and training challenges.

## Teaching objectives

This case is appropriate for students at the honours level (data analytics/science-related course) and Masters of Business Administration (MBA) level degrees. By the end of this case study, students should be able to

- understand the role and importance of business intelligence in the context of skills development
- evaluate the role of data analytics in business intelligence
- develop strategies for leveraging BI to improve skills training programmes.

## Assignment questions

- How does business intelligence aid in skills development within the chemical industry?
- Provide insights into the role and impact of business intelligence in informing strategic planning and decision-making processes for effective skills development.

## Research method

The method of research employed in this case study is a mixed-method approach. Firstly, quantitative data spanning seven years from 2015 to 2022, including WSPs, ATRs and information on levy-paying companies is gathered from CHIETA. This data serves as the foundation for analysing industry trends and identifying skills gaps. Secondly, qualitative data is obtained through interviews with key stakeholders at CHIETA, including the Head of Skills Planning and Development, to gain insights into the complexities and challenges faced in skills sector planning. Additionally, documents and annual reports from CHIETA are collected from online repositories to supplement the information gathered from interviews and quantitative data.

## Additional materials

- CHIETA (2024) Skills sector plan and Trend analysis report: https://www.chieta.org.za/about-us/what-we-do/research-skills-planning/ssp/
- CEDEFOP (2020) online skills forecast dashboard: https://www.cedefop.europa.eu/en/tools/skills-intelligence/sectors

## Key concepts and subject code

The following keywords and the subject code provide a concise summary of the focus of this case study. These keywords and subject codes can help instructors determine whether the case study aligns with the topics covered in their courses and integrate them into their materials accordingly. The instructor should introduce students to the concept of BI and discuss the significance of data analytics, BI and predictive techniques in addressing skills gaps and informing decision-making. In addition, provide real-world examples and case studies showcasing successful implementations of BI systems in the skills development sector (CEDEFOP, 2020).

**Subject code:** Business Intelligence Implementation at CHIETA: Addressing Skills Demands in the Chemical and Manufacturing Industries

**Keywords:** Business Intelligence (BI); Skills Development; Data Analytics; Skills Intelligence; Skills Forecasting; Data Modelling

## Teaching plan and timing

1. ***Preclass Preparation***

- The facilitator should instruct students to read the case study and familiarise themselves with the concept of SETAs and their role in skills development in South Africa.
- Provide additional resources on CHIETA, such as the skills sector plan (SSP) and trend analysis report.
- Encourage students to research and gain insights into the importance of data analytics and BI.

2. ***Case Study Discussion: 30 Min***

Begin the class by summarising the case study and its objectives, emphasising the challenges faced by CHIETA.

Prompt students with questions to guide their analysis, such as.

- What are the challenges faced by CHIETA in addressing the skills demands of the chemical and manufacturing industries?
- How can a BI system help CHIETA improve skills forecasting and predictive techniques?
- What potential benefits can CHIETA and other SETAs gain from implementing a BI system?
- How can data analytics assist in strategic decision-making, funding management and skills intervention effectiveness?

3. ***Practical Application and Recommendations*: *45 Minutes***

- *10 Minutes:* Challenge students to apply the knowledge gained from the case study and develop recommendations for CHIETA or other SETAs considering the implementation of a BI system.
- *10 Minutes:* Instruct students to consider the specific challenges faced by CHIETA and propose strategies for leveraging BI to address those challenges.
- *10 Minutes:* Allow students to present their recommendations and engage in a constructive debate on the feasibility and effectiveness of their proposed strategies.

4. ***Conclusion: 15 Minutes***

- Summarise the key takeaways from the case study and class discussions.
- Highlight the importance of leveraging data analytics and BI in the skills development sector.
- Encourage students to continue exploring the potential applications of BI in addressing skills gaps and the overall efficiency of organisations like CHIETA.

## Analysis of assignment questions

How does BI aid in skills development within the chemical industry? According to Van Der Walt *et al.* (2021), BI solutions are techniques designed to analyse data generated by an organisation's day-to-day operations. For CHIETA, the BI solution was built on this premise, utilising data collected over the last seven years. The collected data undergoes crucial preprocessing to lay the foundation for effective analysis. This step involves exploring the data to understand its contents and prepare it for its intended uses. Python is the preferred tool for preprocessing due to its rich data science packages, including Pandas Profiling for fast exploratory analysis. After preprocessing, data can be securely stored in a database, allowing access through Business intelligence tools like Microsoft PowerBI. This will enable CHIETA to harness BI's capabilities for informed skills development strategies.

### REFERENCES

Alnoukari M and Hanano A (2017) Integration of business intelligence with corporate strategic management. *Journal of Intelligence Studies in Business* **7(2)**: 5–16. https://doi.org/10.37380/jisib.v7i2.235

Bhosale S, Patil R and Medhekar M (2021) *Research Paper on Business Intelligence*, p. 2021.

CEDEFOP (2020). https://www.cedefop.europa.eu/en/tools/skills-intelligence/sectors. Accessed 16 June 2023.

CHIETA (2024) Skills sector plan and Trend analysis report. https://www.chieta.org.za/about-us/what-we-do/research-skills-planning/ssp/. Accessed 16 June 2023.

Niu Y, Ying L, Yang J and Mengqi Bao CBS (2021) Organizational business intelligence and decision making using big data analytics. *Information Processing & Management* **58(6)**. https://doi.org/10.1016/j.ipm.2021.102725

Sivarajah U, Kamal MM, Irani Z and Weerakkody V (2017) Critical analysis of Big Data challenges and analytical methods. *Journal of Business Research*, **70**: 263–286. https://doi.org/10.1016/j.jbusres.2016.08.001

Sivarajah, U., Kamal, M.M., Irani, Z. and Weerakkody, V., 2017. Critical analysis of Big Data challenges and analytical methods. *Journal of business research*, *70*, pp. 263–286.

Van Der Walt A, Gilliland S and Smit I (2021) Data migration in support of business intelligence in a growing organization. *icABCD 2021 – 4th International Conference on Artificial Intelligence, Big Data, Computing, and Data Communication Systems, Proceedings* [Preprint]. https://doi.org/10.1109/icABCD51485.2021.9519357

Wieder B and Ossimitz ML (2015) The impact of business intelligence on the quality of decision making – A mediation model. *Procedia Computer Science* **64**: 1163–1171. https://doi.org/10.1016/j.procs.2015.08.599

Arnesh Telukdarie
ISBN 978-1-83608-365-8
https://doi.org/10.1108/978-1-83608-362-720241003


Chapter 3

# The digital transformation of healthcare maintenance management in Gauteng: a case of Lubuto[1] Hospital

A Telukdarie and J Mwanza

## Introduction

On a late Sunday evening, in the last week of June 2021, Busi[2] and her 4-member senior management team sit around the boardroom table in pin-drop silence, staring blankly at the overhead flatscreen TV. The country's president has just announced, live on national TV, that, with effect from the next day, Monday 28 June 2021, the country is moving to the dreaded alert level 4 lockdown in response to the deadly Delta variant that had become the most dominant strain in the country (Mbhele, 2021). It is now 15 months since the president first declared a COVID-related national state of disaster for South Africa on March 15, 2020, as the COVID-19 pandemic commenced its ferocious grip on the country (Ramaphosa, 2020). In apparent absent-mindedness, each one of them is likely contemplating what this all means for their organisation and musing about the next steps to take. Even although they surely must have seen this coming, it is as if they had all been hoping to be wrong. That, somehow, the president was going to announce a different set of COVID-19 containment measures for the nation. But, in hindsight, that was simply never going to be. The current hospital infrastructure was just too overwhelming for any other announced measure to make sense. Busi and her senior management team in the Manufacturing, Engineering and Related Services Sector Education and Training Authority (merSETA) play a pivotal role in the capacitation of maintenance personnel across all the province's hospitals through the provision of technical training, maintenance planning support and other business solutions. Therefore, the president's message is of extreme importance to them.

After what seems like an eternity, Busi makes a rather loud bang that jolts everyone back to the here and now as she suddenly plants both her palms on the boardroom table, using them to propel her heavy frame upward and onto her feet. After a brief pause during which her eyes scan across everyone in the room, she opens her mouth and begins to speak in her usual authoritative tone. 'Well, this is it colleagues. We have all heard what has been announced. I had asked all of you to come to the office on a Sunday like this, just so that we could all

[1,2] Real names withheld.

hear this important announcement firsthand while we were together in the same room. Let us now call it a night, but tomorrow, we reconvene virtually at 11:00 sharp, and we map the way forward. We need to come up with a robust solution to this challenge of how we are going to arrest and then reverse the trend of declining equipment availability in these hospitals.' After a brief pause, she adds, 'And we need to produce that solution fast. Good night.' With this, she grabs her laptop, loosely dumps it into her bag, zips up the shabby backpack, shuffles out of this rather small boardroom on the eleventh floor of merSETA House in Melville and steps into the thick darkness of the loadshedded Johannesburg night.

## Dire situation

As Busi drives home that night, the dire situation regarding Gauteng's health care maintenance continues to loop inside her head. She reflects on the reality that there simply is not enough medical equipment in service to meet the huge demand caused by the pandemic (McKenzie, 2021). According to the regional health office, there is a serious gap between demand and availability for ICU beds and ventilators in the region's public hospitals (Figure 3.1). Whereas 75% of the region's ICU beds and other emergency care equipment are available at private facilities, the public sector's capacity only accounts for the remaining 25%, and yet it must cater for about 80% of the population (Morrow *et al.*, 2020). Gauteng's COVID-19 patients are hospitalised across 134 facilities, but only 40 of them are public (Alvinejad, 2022). Most hospitals are running critically low on medical oxygen, with five out of the 40 public hospitals in the region even reporting that their oxygen generation plants are completely down for want of service parts. It is estimated that at least 20% of the installed equipment is out of service, exacerbating an already dire situation. Maintenance departments at these hospitals no longer have the luxury of deliberately taking out equipment for routine servicing and, instead, they are now caught up in a seemingly unending cycle of 'firefighting'.

Although the Gauteng MEC for Health, Ms. Mokgethi, claims that hospitals can still cope, it is obvious that hospital facilities are stretched beyond the limit, with many patients having to spend days waiting just for a hospital bed (SA People, 2021). Ever since its onset in the country in early 2020, the pandemic has affected hospital maintenance in various ways, including increased demand for services, staff shortages and supply chain disruptions that impacted the availability of parts. As was the case virtually everywhere else, most health care workers experienced higher workloads and shortage of PPE, social exclusion and stigmatisation, and a lack of proper management, which led to psychological distress and high turnover (Razu *et al.*, 2022; Lalla-Edward *et al.*, 2022). Moreover, the ongoing pandemic placed a tremendous burden on medical equipment across all the public hospitals in her region and has overstretched the capacity of health care workers to operate effectively (Nyasulu and Pandya, 2020).

## The Manufacturing, Engineering and Related Services Sector Education and Training Authority

Busi was the regional director of health services training at merSETA, a role she had thoroughly enjoyed for the past five years and counting. Ever since its establishment through an Act of Parliament way back in 1998, merSETA has been instrumental in the promotion of skills development across manufacturing, engineering and related services sectors within the South African economy (merSETA, 2023). Among other duties, Busi's team was charged with the responsibility of identifying the maintenance needs of medical equipment within all public hospitals in Gauteng, as well as the design, implementation and evaluation of training programs to ensure all the identified needs were met.

As the COVID-19 pandemic caused a high turnover of skilled labor, adversely impacting maintenance planning and execution, several public hospitals within Gauteng reached out to merSETA seeking support. They wanted help in the form of training for maintenance practitioners and managers to close the maintenance planning skills gaps. Their message was always the same: equipment was unable to keep up with the unprecedented rise in demand, and maintenance teams on the ground were so stretched they hardly had time to execute any planned maintenance activities. As a result, equipment failure rates had spiked, and this was now beginning to adversely impact patient outcomes.

## The dilemma

With the usual maintenance strategies all but abandoned and most equipment now operating on a run-to-failure basis, abrupt failures of equipment while in critical use were increasing in frequency, with disastrous consequences for patients. The huge pressures placed on health care infrastructure by the COVID-19 pandemic exacerbated an already dire need to increase the availability of medical equipment at level 1 hospitals. The high demand for ventilators and other lifesaving equipment meant that hospitals could hardly afford the 'luxury' of deliberately taking equipment out of operation for servicing.

Equipment that is constantly in use will eventually fail, often at the most crucial moments. Ironically, the more frequently the equipment failed, the less inclined were operations to allow it to be taken offline for scheduled maintenance. This made it increasingly difficult for maintenance departments to perform preventive maintenance activities, which in turn led to even more equipment failures. As a result, the situation was quickly becoming a vicious cycle of equipment failures and service breakdowns. It was, therefore, critical to determine the appropriate level of maintenance needed for ventilators, defibrillators, oxygen plants and other critical medical equipment, all of which were in high demand.

But how could the right balance be struck between equipment availability for operational activities and planned maintenance? Was there an objective way of adjusting maintenance plans in a way that was sensitive to current demand and the needs of patients? Busi was determined to do all within her power to ensure that this present challenge was resolved, and the direct impact that a solution would have on the thousands of lives suffering from the effects of the pandemic within society motivated her even more.

## Engaging the University of Johannesburg

Busi knew that the solution would require a good measure of innovation, something beyond the routine training of technicians and maintenance engineers they had been offering for quite some time now. After all, extraordinary times call for extraordinary measures. She convinced her team that, in solving this challenging problem, it was prudent for them to engage an external partner to work with. After some preliminary searches, she came across the University of Johannesburg (UJ), which had done quite a lot of work with diverse industries, helping them solve longstanding problems.

They approached UJ and put a straightforward request on the desk of the Johannesburg Business School's (UJ-JBS) research chair. They wanted a sustainable solution that would empower hospital management to understand exactly what was going on with their maintenance operations: where critical resources were required, what the optimal labor allocation to tasks should be, right-sized inventory levels of spare parts, and maintenance plans optimised for frequency of execution. She wanted a solution simple enough that, once implemented, would not require any further oversight from her department or the external partner. A solution that did not depend on the skill and experience of the maintenance planner or engineer.

UJ has collaborated with many industries in the space of digital transformation projects. They were leading the adoption and leveraging of fourth industrial revolution (4IR) tools in transforming businesses and helping organisations find cost-effective solutions to longstanding challenges facing their businesses. So, when this problem was discussed between Busi and the research chair at the UJ-JBS, their ideas converged around exploring how they could leverage 4IR technologies to design a decision-support tool that could help optimise maintenance strategies, reduce downtimes and maximise the life of equipment.

## Development of decision-support tool

Most institutions were already making use of enterprise resource planning systems (ERPs) in their day-to-day operations. Maintenance departments, however, were using a combination of ERP maintenance modules and manual methods for the planning, scheduling and execution of maintenance activities. Maintenance modules were used for activities such as work order generation, work order backlog management, logging of breakdown requests, the requisitioning of spare parts, and general engineering store management. Despite using systems for all these activities, the data generated was not being leveraged other than for reporting purposes. The research team saw that there was an opportunity to utilise this data to generate insights and improve performance. Big maintenance data and the availability of other 4IR tools like simulation and modeling could be combined in useful ways to produce a tool that could serve well as decision support.

The UJ research team used business process modeling to identify key process steps and important parameters in maintenance operations. Then, the research team formulated a multi-equipment maintenance optimisation problem and applied queuing theory and discrete event simulation to simulate the problem. After establishing data collection requirements from the hospital, the UJ research team developed data input forms that they shared with hospital maintenance management and co-filled out. The data collected was about all identified process inputs, process parameters and critical performance indicators for the maintenance department. It included asset registers, details of medical equipment and equipment types, priority rankings, availability, costs, efficiency, labor costs, current mean times between failures (MTBF) and mean times to repair (MTTR) data.

A review of the literature indicated that concerning the development of digital decision-support tools, discrete event simulation (DES), agent-based simulation (ABS) and system dynamics (SD) were the most essential building blocks in the development of automated decision-making for complex systems (Scheidegger *et al.*, 2018). They then developed and proposed a decision support tool that could be used for optimising maintenance workflow in a hospital. The proposed tool combined DES and simulated annealing (SA) in a novel way and could forecast future asset-performance trends and determine the most appropriate interventions. A modified solution space reduction strategy was also applied to improve the tool's search efficiency (Zheng *et al.*, 2021).

## Implementation and outcome

After the tool was developed, it was time to validate it and see if it would produce meaningful output. The UJ researchers ran several simulations using historical input data and compared the program output with actual performance. If there was significant variation from the actual observed results, it would have been concluded that the tool did not accurately or reliably predict performance. As the outcomes of the simulations closely matched the actual performance recorded in the past, it was concluded that the tool was reliable at predicting future maintenance performance. Instead of having to retrain maintenance planners and managers each time new staff are recruited, the developed tool made it possible for them to plan effectively with little skill required on their part. All they had to do was configure the hospital in the digital tool and the tool would indicate how the maintenance strategy was to be set up, including frequency of services, critical spare stock holding policies, manning levels and even equipment prioritisation.

The main 4IR technologies applied in the model's development were simulation and big data analytics. Process logs of maintenance events, including breakdown occurrences, start- and end-times of repairs, service frequencies, times between failures and other parameters, were all collected into a database and later analysed using process mining techniques. The insights obtained therefrom were then used in the optimisation algorithms incorporating DES and SA. The result was a dynamic, adaptable maintenance plan that could be continuously refined based on current performance. Furthermore, the reliance on individual maintenance planning skills and number crunching was rendered redundant. After using historical data to jointly test, debug and validate the decision support tool together with the hospital's maintenance team and merSETA, the UJ team piloted the tool in the maintenance department at Lubuto Hospital. During this trial run, the implementation was closely monitored by the project team, and, thanks to the thorough testing, no major hiccups were experienced. Based on running data obtained over six months, indications were that the solution would deliver significant maintenance cost savings in addition to improving overall equipment availability and service quality, with a reporting of positive patient outcomes likely to ensue. Over that period, the availability of ventilators improved by 15% and the trend of MTBF data was beginning to show signs of reversal. The number of breakdowns decreased month on month. Although spare consumption went up, the reduction in the cost of downtime and labour costs led to lower overall maintenance costs.

## Future outlook

Was it too early to make conclusive inferences from the tool's performance from just this one implementation? Busi was content with the outcome that her team's collaboration with the UJ-JBS had produced and with the positive impact that the improvement in equipment availability was now having on patient outcomes. However, it remains to be seen if the tool will continue to offer value in the aftermath of the COVID 19 pandemic when perhaps the focus on maintenance activities will be diluted back to pre-pandemic levels. Busi also wondered if terotechnologists across the health sector and beyond would embrace this quick-to-implement and cost-effective decision-support tool. Or perhaps the tool's impact would expand beyond just health care maintenance and find even broader applications? 'Time will tell' she finally concluded to herself, deep in the recesses of her ever-scheming mind.

## Exhibits

Figure 3.1 ICU beds at private against public health care facilities (author's own)

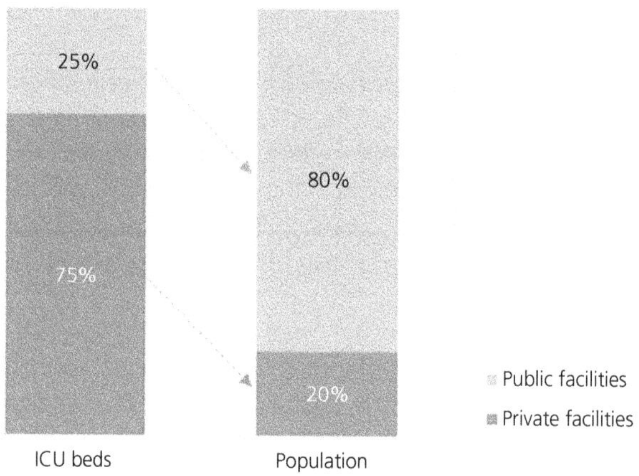


## Teaching notes

### Synopsis

This case study documents Busi's efforts as the research director at merSETA to create an intelligent, cost-effective solution for optimising maintenance in Gauteng hospitals during the height of the COVID-19 pandemic in 2020 and 2021. Due to the overwhelming demand for medical equipment and inefficient maintenance practices, the system was at risk of collapsing. It was crucial to find innovative and affordable ways to maintain equipment performance and reliability to continue fighting the pandemic effectively. The study details the research team's approach, which involved reducing the business problem to an optimisation objective and using simulation techniques and optimisation meta-heuristics to create a decision-support tool for developing a reliable maintenance strategy. The case study's success in reducing downtime and repair times highlights how digital technologies can transform businesses and have a positive impact.

### Research methods

The authors spoke to the research director using a structured question set to understand the following.

- The significant details of the challenge that the institution was facing.
- The desired outcomes that the intervention would bring.

- The vision, mission and strategic objectives of the organisation and to understand how solving this challenge would help the organisation towards achieving them.

As this was quantitative research, a hybrid simulation modeling approach combining elements of discrete event simulation and simulated-annealing-based optimisation was adopted in developing the decision-support tool. Equipment and cost data provided by the hospital's maintenance and finance departments were used to test and validate the tool. This data included equipment lists, maintenance schedules, repair records, parts inventory and cost data, as well as artisan rates. Relevant historical data was collected covering 24 months up to December 2020. The data was cleaned, normalised and stored in a MySQL database. The digital tool's logic was implemented in Python using the SimPy simulation library.

## Teaching objectives

This case study is intended for use by postgraduate students studying business management courses. The specific learning objectives for this case study are as follows.

- To give students an understanding of the business research process followed by the JBS research team and how the same approach can be applied to solve business problems in other contexts. Students are expected to develop the ability to conduct business research processes when tackling business problems with multiple variables and factors.
- To equip students with the tools and skills that facilitate objective decision-making related to operations issues that may be encountered within a health care maintenance environment. Business management students are expected to develop the skill to approach decision-making in an objective, scientific manner.
- To expose students to various models of decision-making available in the literature and equip them with skills to select and apply the most appropriate techniques and practices (such as frameworks, protocols, checklists, or policies) for developing effective management tools for maintenance-related problems.

## Assignment questions

Question 1: Given the digital technology tools now available, how would you approach the maintenance optimisation problem for a health care institution situated in a resource-constrained environment?

Question 2: List any three optimisation techniques that may be applied in the development of a data-driven decision-support tool and briefly explain how each of the techniques you have listed works.

Question 3: In this case study, which would you say are the key indicators of effective maintenance management that the research team focused on optimising?

Question 4: Identify and briefly discuss a different business context, other than health care maintenance, in which the approach used in this case study would be appropriate.

## Suggested readings

The sources listed below give the student additional reading material related to the concepts and techniques used in this case study.


- Alrabghi A and Tiwari A (2016) A novel approach for modeling complex maintenance systems using discrete event simulation. *Reliability Engineering & System Safety* **154**: 160–170. https://doi.org/10.1016/j.ress.2016.06.003
- Fotovatfard A and Heravi G (2021) Identifying key performance indicators for health care facilities maintenance. *Journal of Building Engineering* **42**: 102838. https://doi.org/10.1016/j.jobe.2021.102838
- Sánchez-Barroso G and Sanz-Calcedo JG (2019) Application of predictive maintenance in hospital heating, ventilation and air conditioning facilities. *Emerging Science Journal* **3(5)**: 337–343. http://dx.doi.org/10.28991/esj-2019-01196
- Warrington L, Jones JA, and Davis N (2002) Modeling of maintenance, within discrete event simulation. *Annual Reliability and Maintainability Symposium. 2002 Proceedings (Cat. No.02CH37318)* (pp. 260–265). https://doi.org/10.1109/RAMS.2002.981652
- Yousefli Z, Nasiri F, and Moselhi O (2020) Maintenance workflow management in hospitals: An automated multiagent facility management system. *Journal of Building Engineering* **32**: 101431. https://doi.org/10.1016/j.jobe.2020.101431


## Keywords and subject code

Decision-support, Digital Transformation, Maintenance, Modeling, Optimisation, and Performance Management

## Teaching plan and timing

It is recommended that this case be given to students as a pre-read before going through the material in a classroom setting, as some of the concepts introduced may be rather complex and require background reading. Additional reading materials are included in the teaching note. The case study may be read out aloud in a class, delivered online or conducted in a hybrid manner, but with appropriate pauses for reflection and discussion after each section is covered. It is perhaps best to give the students a 15-minute slot at the end for group discussion where they will recap the main takeaways to each other to enhance their overall understanding of the concepts involved.

All in all, this material will require about 90 minutes, as follows.

- *5 min*: Topic presentation, reading of synopsis and discussion of the purpose of the study.
- *30 min*: Section-by-section reading and discussion of the case.
- *10 min*: Recap and review of learning objectives.
- *15 min*: Group discussions and feedback.
- *15 min*: Students working independently to address the questions.
- *15 min*: Review of students' answers.

## Analysis of assignment questions

***Answer to Assignment Question 1***: The student's answer should demonstrate a good grasp of the principles and theories of systems optimisation. The answer should also demonstrate familiarity with 4IR digital tools, particularly those applicable in developing decision-support tools for maintenance management within a resource-constrained environment. The student's answer must indicate a bias towards tools and methodologies that most readily address the specific challenges faced by low-resourced economies, including limited equipment availability, a general scarcity of spare parts and a shortage of skilled maintenance resources. Finally, the answer should also propose data collection mechanisms that can gather relevant data in real-time to inform maintenance decisions on an ongoing basis.

***Answer to Assignment Question 2***: The student's answer should offer a concise explanation of any three optimisation techniques that may be applied, in combination, in the development of a data-driven decision support tool. It is expected that the techniques mentioned in the student's answer may include any of the following: discrete event simulation (DES), simulated annealing (SA), agent-based modeling (ABM), and any of the various machine learning algorithms available, such as decision trees, random forest, k-nearest neighbors (KNN) and neural networks.

***Answer to Assignment Question 3***: The student's answer to this question should highlight the key performance indicators of an effective maintenance program that the research team focused on. It is expected that the answer will mention any or all the following indicators: maintenance costs, mean time between failures (MTBF), mean time to repair (MTTR), equipment availability, manning levels, and labor costs.

***Answer to Assignment Question 4***: The student should identify a different business context in which the approach used in this case study would be applicable. The identified business context could be fleet optimisation in logistics, process flow optimisation in service industries such as banks or hospitals, or maintenance management within small and medium-sized enterprises (SMEs). The student should provide a very brief explanation of how this same approach could be adapted for application in the selected business context.

### Key learnings

The instructor can conclude the case discussion with a summary of how the case demonstrates two to three key learnings – for example, the centrality of cash flow management in organisational sustainability or the importance of responding to a wide range of stakeholders in different ways, as well as providing the students with a model or framework given as an exhibit that they could apply in the future to solve other problems they may face in their careers. These must be stated clearly here in about 150 words.

## Conclusion

This case study demonstrates how merSETA collaborated with JU-JBS's research team to address a business problem using digital tools from Industry 4.0. Although there was only a limited time in which to understand and solve the problem due to demand pressures during the peak of the pandemic, the team successfully developed a solution in the form of a data-driven decision-support tool to aid in the management of critical hospital equipment, including ventilators, blood gas analyzers, infusion pumps and ambulances. The lessons drawn from this case study have practical applications for improving maintenance performance in other fields of business, as they offer information that is of relevance to maintenance management practitioners in general. The case is also academically significant as it proposes a unique approach to addressing this type of optimisation problem through seamless integration of discrete event simulation, optimisation metaheuristics and continuous optimisation forecasts based on real-time operational metrics. These novel contributions have the potential to advance theory in the field of systems optimisation, particularly within the maintenance management context.

## REFERENCES


The Citizen (2021) Level 4 lockdown for 2 weeks: Alcohol sales and public gatherings banned. *The Citizen*, 27 Jun. https://www.citizen.co.za/news/level-4-lockdown-for-2-weeks-alcoholsales-and-public-gatherings-banned/ (accessed 03/05/2023).

Ramaphosa (2020) Statement by President Cyril Ramaphosa on measures to combat the COVID-19 epidemic. https://www.thepresidency.gov.za/press-statements/statement-presidentcyril-ramaphosa-measures-combat-covid-19-epidemic (accessed 03/05/2023).

McKenzie D (2021) Southern Africa hoped it was through the worst of Covid-19. Then the Delta variant arrived. *CNN*, 09 Jul. https://www.cnn.com/2021/07/09/africa/southern-africa-covid-delta-intlcmd/index.html (accessed 03/05/2023).

Morrow BM, Gopalan PD, Joubert I, Paruk F and Pope A (2020) Critical care triage during the COVID-19 pandemic in South Africa: A constitutional imperative. *South African Medical Journal* **110(12)**, Art. no. 12. https://doi.org/10.7196/SAMJ. 2020.v110i12.15411

M. Alavinejad, Mellado B, Asgary A *et al.* (2022, Nov.) Management of hospital beds and ventilators in the Gauteng province, South Africa, during the COVID-19 pandemic. *PLOS Global Public Health*, **2(11)**: e0001113. https://doi.org/10.1371/journal.pgph.0001113

SA People (2021) Gauteng hospital shortages endangering lives with 'Tsunami of New Cases'. *SA People - Worldwide South African News*, 21 Jun. https://www.sapeople.com/2021/06/21/gauteng-hospitalshortages-endangering-lives-with-tsunami-of-new-cases/ (accessed 03/06/2023).

Razu SR, Yazmin T, Arif TB *et al.* (2021) Challenges faced by healthcare professionals during the COVID-19 pandemic: a qualitative inquiry from Bangladesh. *PubMed* **9**. https://pubmed.ncbi.nlm.nih.gov/34447734/ (accessed 27/04/2023).

Lalla-Edward ST, Mosam A, Hove J *et al.* (2022) Essential health services delivery in South Africa during COVID-19: Community and healthcare worker perspectives. *Frontiers in Public Health*, **10**: 992481. https://doi.org/10.3389/fpubh.2022.992481

Nyasulu J and Pandya H (2020) The effects of coronavirus disease 2019 pandemic on the South African health system: A call to maintain essential health services. *African Journal of Primary Health & Family Medicine* **12(1)**. https://doi.org/10.4102/phcfm.v12i1.2480

MerSETA. https://www.merseta.org.za/about-us/overview/ (accessed 03/06/2023).

Scheidegger A, Pereira TF, de Oliveira M, Banerjee A and Montevechi JAB (2018) An introductory guide for hybrid simulation modelers on the primary simulation methods in industrial engineering identified through a systematic review of the literature. *Computers & Industrial Engineering* **124**: 474–492. https://doi.org/10.1016/j.cie.2018.07.046

Zheng Q, Feng B-W, Chang H-C and Liu Z-Y (2021) Dynamic space reduction optimization framework and its application in hull form optimization. *Applied Ocean Research* **114**: 102812. https://doi.org/10.1016/j.apor.2021.102812

Arnesh Telukdarie
ISBN 978-1-83608-365-8
https://doi.org/10.1108/978-1-83608-362-720241004


Chapter 4

# Digital transformation for steel: a case of BBT steel manufacturers

A Telukdarie and BG Mwanza

## Role of steel mining in South African development

The steel industry assists in the development of South African infrastructure programs. Essentially, it has the ability and potential to finance the pivotal reindustrialization agenda in South Africa. The South African Iron and Steel Institute (SASTI) emphasizes the criticality of addressing the three sustainability pillars within the steel industry. This is crucial due to steel's pivotal role in a nation's economy and its contribution to national development agendas.[1,2] The importance of sustainability in the steel industry cannot be overemphasized, especially since global steel demand has increased by 700% since 1950 and it is projected to increase by 150% by 2050.[3] The steel industry has great influence in the following spheres of the economy: *Manufacturing*: This industry utilizes steel equipment and machinery for the construction of customers' goods. *Reusability:* The steel recycling and reusing industry contributes to the reduction of environmental impacts by reducing steel production. *Employment:* Throughout the value chain of the steel industry, jobs have been created from raw material extraction, production and utilization. *Wealth:* Funds from the corporate entities dealing in steel are obtained through government taxes and legislation in the steel industry. *Community:* The livelihoods of many steel communities are enhanced through job creation, education and experiences. *Commerce:* Steel supply chains create financial benefits globally.

## Michael and the integration challenge

BBT Steel Manufacturers was established in 2005 and is an Altx-listed South African-based company. It has strong ties to southern African markets. BBT Steel Manufacturers vision is to be a sub-Saharan partner in steel solutions in Africa. Our company has upgraded its technology, leading to the integration of its manufacturing and business systems to meet its forecasted demand. With the growth in market share in the African region, it has managed to purchase Zambia and Congo businesses as well as convert all its subsidiaries within the group of wholly owned subsidiaries. Our product range is made up of mild steel in all its forms and we also offer a very extensive range of tubing products to the building and manufacturing industries. Our company is committed to providing corporate responsibility-related support to the communities in which we operate.

Sarah Michael, the managing director of BBT steel manufacturers based in South Africa, analyzed the steel demand and noted that the demand is projected to increase until the year 2050. Michael was excited about the projected increase, but she also noted that there are challenges that BBT steel manufacturers are likely to face in forecasting the demand. The main challenge faced by BBT steel manufacturers is the disconnect that exists between shop floor systems such as manufacturing execution systems (MES) and business systems such as enterprise resource planning (ERP). As a result of the disconnect, BBT has challenges managing and interpreting its data because the devices that transmit or store data have limited data sets or there is a lack of standardization. Consequently, Michael realized that without the integration of manufacturing and business systems, it would be challenging to forecast demand and later make real-time decisions.

As a way of resolving the challenge, Michael decided to engage the University of Johannesburg, Johannesburg Business School (UJ-JBS) as a consultant. As a strategic approach, it was necessary to constitute a team from both partners. The team that was constituted had six members, 3 from each organization. The manufacturing and business systems for BBT steel manufacturers required integration and therefore the team developed two objectives. The first object investigated the technologies and infrastructure that steel companies in developed economies employed. The second objective established how to leverage the investments to attain smart enterprises that would eventually manage current and future steel demands efficiently.

To achieve the task at hand, Michael and the lead researcher from UJ-JBS proposed the application of Industry 4.0 technologies. They argued that Industry 4.0 simplifies the implementation of complex systems in a logical and structured manner. While the application of Industry 4.0 was agreed upon, the expected results to resolving the challenge were to develop an Industry 4.0 framework that cannot only be applicable at BBT steel manufacturers but also in steel manufacturing companies of developing economics.

## Leveraging Industry 4.0 technologies for integration

The fourth industrial revolution describes the approach to smart manufacturing using technologies in the industry 4.0 framework. To leverage Industry 4.0 technologies, BBT steel manufacturers and UJ-JBS considered the integration of manufacturing and business systems based on the benefits that specific Industry 4.0 technologies present. In this case, not all the technologies are applied.

### Cyber-physical systems (CPS) and industrial internet of things (IIoT)

The ability of smart devices to receive and transmit data packets from themselves and other devices is referred to as the internet of things (IoT) (Stock and Seliger, 2016), while the communication and collaboration of smart devices in the industrial environment are defined as the industrial internet of things (IIoT). It was necessary to apply the definition in terms of how intelligent devices for BBT will be enhanced to comply with the protocols and conditions of the manufacturing operation. Since business and manufacturing systems require integration, IIoT provides an enabling environment for Industry 4.0 integration layers to fulfill a real-time data flow. Three integration layers from the Industry 4.0 framework were considered for application: *(i)* vertical integration, which enables the flow of data from the manufacturing system to the ERP; *(ii)* end-to-end integration, which manages and maintains the flow of product data through product life cycle management; and *(iii)* horizontal integration, which enables the sharing of information between the supply chains of organizations. The consideration of these integration layers of Industry 4.0 was key to resolving the demand for forecasting challenges.

CPS gives insights into manufacturing operations to permit a company to improve and act on any challenges it may anticipate (Mousa *et al.*, 2016). In this case, Michael and her team are glad to leverage this aspect presented by CPS technologies, as these technologies will be able to collate data from IIoT devices, present notifications and contextualize data from manufacturing operations with business data to give a holistic view of BBT steel manufacturers with regards to their produced stock, completed work orders and work-in-progress. Further, the CPS will enable BBT steel manufacturers to access data on steel availability within the supply chains to meet customer demand.

## Big data

In the steel industry, a combination of automated data and manual data entry is utilized to manage the companies. However, validating and contextualizing data from several sources presents a big challenge. Nevertheless, IIoT technologies provide a platform to share any available data in real-time, while big data platforms store information in a legible and structured manner (Pandia *et al.*, 2016). In this case, big data platforms will enable BBT steel manufacturers to have a single source of validated information that is first checked and verified before storage. It will enable BBT steel manufacturers to predict and forecast the future.

## Smart manufacturing

The IoT, people and devices is referred to as a smart plant and its characteristics include the following (Olalekan *et al.*, 2016).

- **Maintenance** is concerned with implementing and managing maintenance strategies that attune to the information communicated from the devices to reduce downtime.
- **Analytics** is concerned with providing visibility of data acquired by way of CPS and IIoT devices to define strategies for improving operational efficiency using collated data.
- **Monitoring** is concerned with the real-time visibility of operational devices to ensure the devices are maintained and managed remotely using IIoT.
- **Process optimization** is concerned with managing and tracking implemented improvements in the company. This characteristic of smart manufacturing enables companies to benchmark whether changes result in improved operations.
- **Utility management** is concerned with aligning energy supply and demand to make sure energy peaks are controlled.
- **Automation** is concerned with automating repetitive tasks for quality and efficiency improvement.

- **Flexibility** is the ability of operations to adapt to product specifications or quality changes.
- **Supply chains** are concerned with suppliers, customers and operations visibility to ensure proactive problem resolution.

Michael and her team considered the integration of smart manufacturing technology because its characteristics provide an opportunity for the company to resolve demand forecast challenges. Further, the characteristics of smart manufacturing are aligned with big data, IIoT and CPS, which will enhance the development of an Industry 4.0 framework.

## Developing an Industry 4.0 framework for steel manufacturing companies

The approach to developing an Industry 4.0 framework was based on the systems development life cycle (SDLC) process. This process began with establishing the current infrastructure and technologies that are used in South African steel companies. To begin with, interviews and workshops were held with several stakeholders that operate at different levels within BBT Steel Manufacturers. The future and current state of BBT Steel Manufacturers was explored and recorded, which resulted in many versions of business requirements specification (BRS) documents. The final version of BRS was used to make system design artifacts as the functional design specification (FDS) and technical design specification (TDS). The design documents gave an architecture of a completely connected enterprise that embodies the Industry 4.0 framework (see Figure 4.1).

After the requirements gathering and analysis phases, Michael and her team learned that the company utilizes 116 independent systems daily. They also learned that there were duplicate systems, such as weighbridge solutions, which are not standardized, and that the company does not have a central data repository covering the several sites. System classification was also lacking and, in this case, ERP and MES were not classified, which resulted in the utilization of both systems as either MES or ERP functionalities.

After identifying these challenges, Michael and her team developed an Industry 4.0 framework that works for both BBT steel manufacturers and other steel companies (see Figure 4.1). To begin with, Michael and her team performed the systems review process to remove the duplicated systems as a strategy to consolidate the system results. This activity resulted in 50 systems from the 116 that were identified earlier. The team was excited to notice a reduction in the duplicated systems, but the reduction of the systems did not in itself come without challenges. The challenge was in the definition of a business architecture in which systems and business processes would repose.

## Benefits of implementing the Industry 4.0 framework at BBT steel

The developed architecture is ready for implementation but is the team ready to implement it? That was the question they asked each other. However, in this challenge, the scope was to develop an Industry 4.0 framework that could apply to the steel industry and then apply it to observe the changes in the operations. In this case, Michael and her team did not implement it at BBT Steel Manufacturers, but they were able to present what would be achieved from the developed architecture.

IIoT devices would relay real-time information because of the embedded actuators and sensors. Information from machines and sensors is propagated to historians, SCADA and PLCs in a unified structure. As a result of this, information is consolidated into MES systems and made available to specialized applications such as ERP or incident management for equipment monitoring, process optimization and maintenance management. IIoT devices present information to the MES by utilizing the integration bus that manages information transfer between the ERP and MES systems. The information is relayed to the ERP for safety and human resource management. The designed architecture enables end-to-end integration. For instance, product design stages are integrated into the architecture because data is transferred from IoT devices to ERP and MES systems. Horizontal integration is also enabled through the designed architecture. In this case, customers, operations and suppliers are connected.

The architecture utilizes big data technologies that consist of several layers of information in a company. In this case, BBT steel manufacturers, stored information can be transferred through the IIoT sensors. Concerning smart manufacturing, the architecture realized several smart factor characteristics such as supply chain, process optimization, automation, monitoring, maintenance flexibility and analytics. Supply chain activities for managing the availability of raw materials and ensuring customer demand are sanctioned through the IIoT. BBT steel manufacturers' processes will be optimized because of the existence of information from customers and suppliers. Repetitive tasks are automated and data is captured and retrieved in real time from the source systems. IIoT presents an opportunity for the extraction of real-time data from devices.

The architecture design provides flexibility through smart manufacturing. Product changeovers are enhanced because of visibility on time requirements to effect changes because of end-to-end integration. Finally, analytics data received and consolidated through vertical, horizontal and end-to-end integration provides the required insights. The benefits to be realized from the implementation of the designed architecture are cardinal to BBT Steel Manufacturers' demand forecasting challenges.

## Moving forward

The implementation of the Industry 4.0 framework at BBT steel manufacturers can promote standardization, reusability of technologies and data centralization. BBT steel manufacturers have taken advantage of the long-term benefits that come with the implementation of Industry 4.0 frameworks, as they have been able to access insights on challenges within their operations and improve their integrated systems. Michael and her team acknowledge that the integration of the manufacturing and business systems does present some challenges, but they are prepared to tackle them as the operations continue. They are excited that BBT steel manufacturers have been presented with an opportunity to tap into the projected demand for steel with foresight.

# Exhibits

Figure 4.1 Proposed steel industry architecture (author's own)

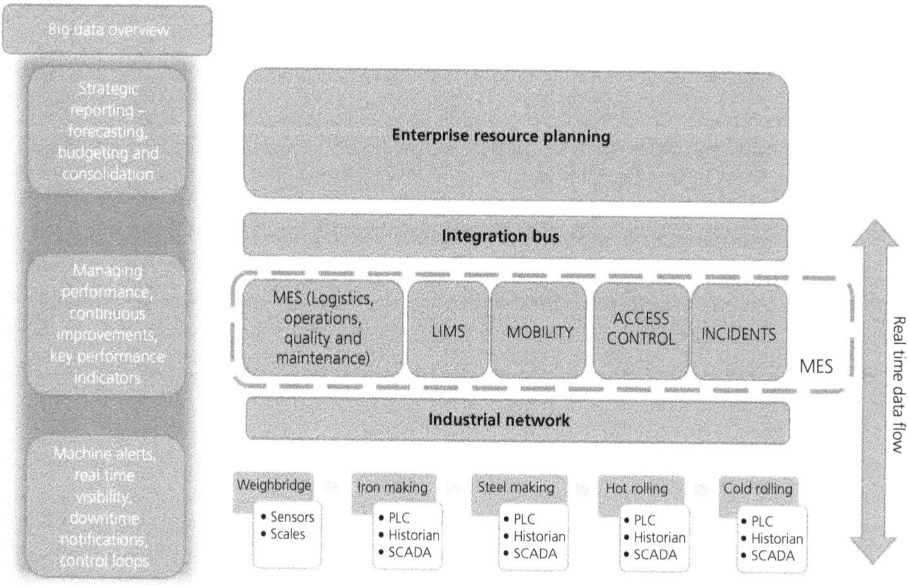


# Teaching notes

## Synopsis

The Industry 4.0 implementation case is the story of a South African steel manufacturing company that developed an Industry 4.0 framework to integrate its manufacturing and business systems. Sarah Michael, the Managing Director (MD) of BBT steel manufacturers, reported that steel demand will increase until the year 2050. For the company to meet its forecasted demand each year, it should integrate manufacturing and business systems. Therefore, BBT steel manufacturers had to partner with the UJ-JBS to integrate the manufacturing and business systems. Shortly after this, Michael and the researcher from UJ-JBS built a team of experts in systems integration and tasked them to conduct the following tasks: *(i)* investigate the current technologies and investments that steel companies in developed economies employ and *(ii)* establish how to leverage the existing investments to attain smart enterprises that would eventually manage current and future steel demands efficiently. To achieve this task, Michael proposed that the team use Industry 4.0 technologies as they simplify the implementation of complex systems in a logical and structured manner. The expectation from the team is the development of an Industry 4.0 framework that companies in the steel industry can implement.

## Teaching objectives

This case is planned for use in business systems, manufacturing systems and Industry 4.0 technologies related to MBA, executive education and undergraduate courses. With consideration of the nature of the stated teaching objectives, this case is suitable for the second stage of the program postgraduate. The case consists of five pedagogical objectives.

- To examine the current manufacturing systems (manufacturing execution system) and business systems (enterprise resource planning) utilized by steel companies portrayed in this case.
- To understand how to leverage these existing manufacturing and business systems to achieve a smart enterprise as depicted in the story.
- To establish the appropriate Industry 4.0 technologies for implementation in a logical and structured manner.
- To develop an Industry 4.0 framework for implementation in any steel-related industry.
- To demonstrate the path for implementing the industry 4.0 framework as depicted in the case.

### Research methods

The implementation of an Industry 4.0 framework, as portrayed in this case study, is based on the requirement engineering and system design steps of the systems development life cycle (SDLC) process. The procedure begins with establishing the current infrastructure and technologies used by steel companies in South Africa.

Interviews and workshops were conducted with many relevant stakeholders working at disparate levels at BBT Steel Company. The current and future conditions of BBT Steel Company were explored and documented, resulting in numerous types of business requirements specification (BRS) documents.

The final type of BRS was used to make system design artifacts such as the functional design specification (FDS) and technical design specification (TDS). The design documents demonstrate the architecture of a completely connected enterprise that embodies the Industry 4.0 framework.

## Keywords and subject code

Many important ideas have been addressed in this case and they include manufacturing systems, Industry 4.0, implementation, integration of business systems, Industry 4.0 progression, big Industry 4.0 technologies, management, framework development and smart enterprise.

There is no right or wrong solution to addressing the identified problem in this case. Students are encouraged to use each concept tackled in this case and establish the solutions that BBT Steel Manufacturers used.

### Assignment questions

- What role does the integration of business and manufacturing systems play in achieving financial sustainability?
- What is the role of Industry 4.0 in the integration of business and manufacturing systems?
- What is the procedure for implementing or adopting new technology in the steel industry?

## Additional materials

- Video: Industry 4.0 and the Transformation of the Steel Company; https://www.youtube.com/watch?v=XPhWcbFW5mM. This is a 3-min video to demonstrate the transformation of the steel industry through Industry 4.0.
- Video: Application of the Industry 4.0 Concept in a Steel Company; https://www.youtube.com/watch?v=XPhWcbFW5mM. This is a 2:45-min video to demonstrate the application of Industry 4.0 in the steel industry.

## Suggested readings

The following materials are recommended articles for enhancing understanding of the case.


- Dalenogare LS, Benitez GB, Ayala NF and Frank, AG (2018) The expected contribution of Industry 4.0 technologies for industrial performance. *International Journal of Production Economics* **204**: 383–394.
- Frank AG, Dalenogare LS and Ayala NF (2019) Industry 4.0 technologies: implementation patterns in manufacturing companies. *International Journal of Production Economics* **210**: 15–26.
- Gajdzik B and Wolniak R (2022) Framework for R&D&I activities in the steel industry in popularizing the Idea of Industry 4.0. *Journal of Open Innovation: Technology, Market, and Complexity* **8(3)**: 133.
- Qin J, Liu Y and Grosvenor R (2016) A categorical framework of manufacturing for Industry 4.0 and beyond. *Procedia CIRP* **52**: 173–178.
- Tsebe A (2022) Business cycles and growth of South African Steel Manufacturing Industry. *Journal of Economics and Behavioral Studies* **14(1)**: 6–22.


## Teaching plan and timing

The case can be distributed to the class as prereading. It can be used to demonstrate the integration of business and manufacturing systems as a road map for the digital transformation of steel manufacturing industries.

The timing for a 2-h class is as follows.

- *5 min*: Introduction
- *20 min*: Analyzing the current state of steel manufacturing industries in South Africa.
- *25 min*: Identification of business systems and manufacturing systems in steel manufacturing.
- *25 min*: Discussion of Industry 4.0 technologies and their adoption at BBT steel manufacturers.
- *30 min*: Development of the Industry 4.0 framework and its implementation at BBT steel manufacturers.
- *10 min*: Key learning.

## Analysis of assignment questions

**Q1. What role does the integration of business and manufacturing systems play in achieving financial sustainability?**

BBT steel manufacturers have realized that they are operating in a competitive industry where they must provide accurate demand forecasts to match the supply. To achieve this, they needed to integrate the business and manufacturing systems for them to achieve financial sustainability. Further, business and manufacturing systems integration contributes critically to achieving financial sustainability through:

- Cost optimization: By integrating business and manufacturing systems, organizations can gain better visibility and control over their production processes, supply chain and inventory management. This enables them to identify and eliminate inefficiencies, reduce waste, streamline operations and optimize costs throughout the manufacturing lifecycle. Improved cost management directly impacts financial sustainability by maximizing profitability and minimizing expenses.
- Demand and capacity planning: Integration allows organizations to align their manufacturing capabilities with market demand. By integrating business systems such as sales, marketing and forecasting with manufacturing systems, companies can obtain real-time data on customer demand, market trends and sales forecasts. This enables effective demand planning, production scheduling and capacity utilization, avoiding overproduction or underutilization of resources. Matching production levels with demand helps reduce inventory costs, avoid stock outs and optimize resource allocation.
- Supply chain efficiency: Integrating business and manufacturing systems facilitates collaboration and coordination with suppliers and partners. With real-time visibility into supply chain data, organizations can optimize procurement, reduce lead times, manage supplier relationships and negotiate better terms. Streamlining the supply chain improves cost efficiency, minimizes disruptions, enhances product availability and strengthens the overall financial sustainability of the organization.
- Quality management: Integration enables organizations to implement robust quality management systems across the entire manufacturing process. By linking business systems with manufacturing systems, organizations can capture and analyze quality data, implement quality control measures and track product performance. This helps identify and rectify quality issues promptly, reduce defects, minimize rework and improve customer satisfaction. Enhanced quality management contributes to financial sustainability by reducing warranty costs, enhancing brand reputation and increasing customer loyalty.
- Data-driven decision making: Integration of business and manufacturing systems provides access to real-time data and analytics. This enables organizations to make informed decisions based on accurate and up-to-date information. Data-driven decision-making supports financial sustainability by optimizing resource allocation, identifying growth opportunities, reducing risks and enabling proactive management of production and business processes.
- Flexibility and agility: Integration allows organizations to respond quickly and effectively to market changes, customer demands and competitive pressures. By

connecting business and manufacturing systems, companies can adapt production plans, adjust inventory levels and modify strategies promptly. This flexibility helps organizations navigate market uncertainties, minimize excess inventory, avoid obsolescence and maintain financial stability.
- Continuous improvement: Integration promotes a culture of continuous improvement by facilitating the exchange of information and feedback between business and manufacturing systems. Organizations can leverage performance data to identify areas for improvement, implement process enhancements and drive innovation across the value chain. Continuous improvement enhances operational efficiency, reduces costs and fosters long-term financial sustainability.

After integrating the business and manufacturing systems, BBT steel manufacturers were able to optimize costs, align production with demand, enhance supply chain efficiency, improve quality management, make data-driven decisions and adapt to market dynamics. The enablement collectively contributed to BBT steel manufacturers achieving financial sustainability by maximizing profitability, minimizing waste and costs, and enhancing operational efficiency.

**Q2. What is the role of Industry 4.0 in the integration of business and manufacturing systems?**

Industry 4.0, also known as the fourth industrial revolution, refers to the integration of advanced digital technologies and automation in manufacturing processes. Industry 4.0 plays a pivotal role in the integration of business and manufacturing systems by enabling seamless connectivity, data exchange and intelligent decision-making. Here are some key aspects of Industry 4.0 and its role in integration.

- Connectivity and interoperability: Industry 4.0 leverages technologies such as the Internet of Things (IoT), cloud computing and cyber-physical systems to establish connectivity and interoperability between various business and manufacturing systems. This allows for real-time data sharing, communication and integration across different stages of the value chain, including production, supply chain and customer relationship management.
- Data analytics and insights: Industry 4.0 enables the collection and analysis of vast amounts of data from different sources within the manufacturing process. Advanced analytics tools and techniques, such as artificial intelligence (AI) and machine learning (ML), are used to extract valuable insights, identify patterns, optimize operations and support data-driven decision-making. This enables organizations to make informed decisions based on accurate, timely and actionable information.
- Smart manufacturing and automation: Industry 4.0 incorporates technologies like robotics, autonomous systems and automation to create smart manufacturing environments. These technologies enable the seamless integration of machines, equipment and systems, resulting in increased productivity, efficiency and flexibility. Smart manufacturing systems can self-monitor, self-optimize and self-adjust, improving overall operational performance and responsiveness.

- Predictive maintenance: Industry 4.0 facilitates the implementation of predictive maintenance practices by leveraging sensor data, real-time monitoring and analytics. By continuously monitoring the condition of machines and equipment, organizations can detect potential failures or maintenance needs in advance. This proactive approach helps reduce unplanned downtime, optimize maintenance schedules and minimize costs associated with equipment breakdowns.
- Supply chain integration: Industry 4.0 enables the integration of business systems with suppliers, logistics providers and customers. Through technologies like IoT, RFID and blockchain, organizations can achieve end-to-end visibility and traceability in the supply chain. This integration facilitates efficient inventory management, optimized logistics, enhanced demand forecasting and improved collaboration with supply chain partners.
- Customization and personalization: Industry 4.0 enables the implementation of flexible and customizable manufacturing processes. By leveraging technologies like 3D printing, digital modeling and data-driven customization, organizations can efficiently produce customized products to meet individual customer preferences. This integration of customer demands into the manufacturing process helps enhance customer satisfaction, increase market responsiveness and drive revenue growth.
- Agile decision-making: Industry 4.0 provides real-time data and analytics capabilities that support agile decision-making. By connecting business systems with manufacturing systems, organizations can monitor key performance indicators, track production progress and respond quickly to changing market conditions. This agility allows for faster decision-making, adaptive production planning and improved operational efficiency.

Overall, Industry 4.0 plays a critical role in the integration of business and manufacturing systems by enabling connectivity, data exchange, automation and advanced analytics. It enhances the visibility, efficiency and agility of manufacturing processes, supporting better decision-making, optimized operations and improved financial performance. By embracing Industry 4.0 technologies, organizations can unlock new opportunities for growth, competitiveness and sustainability in the modern manufacturing landscape.

**Q3. What is the procedure for implementing or adopting new technology in the steel industry?**

Implementing or adopting new technology in the steel industry involves a systematic procedure to ensure a smooth transition and successful integration. While specific steps may vary depending on the nature of the technology and the organization, here is a general procedure that can be followed.

- Identify technology needs: Determine the specific technology requirements and needs of the steel industry. This can be done by conducting a comprehensive analysis of the current state of operations, identifying pain points and assessing areas where technology can bring improvements and efficiencies.
- Research and evaluation: Conduct thorough research to identify potential technologies that address the identified needs. Evaluate different options based on factors such

as functionality, compatibility, scalability, cost and potential benefits. Engage with technology vendors, attend industry conferences and consult with experts to gather information and insights.

- Business case development: Develop a business case that outlines the potential benefits, costs and return on investment (ROI) of the proposed technology. This includes quantifying the financial impact, operational improvements and competitive advantages it can bring to the organization. The business case should also consider the risks, challenges and implementation requirements.
- Stakeholder alignment: Engage and align stakeholders within the organization, including top management, operational teams, the IT department and relevant departments. Communicate the benefits of technology and address any concerns or resistance. Ensure that all stakeholders understand and support the goals and objectives of the technology implementation.
- Pilot testing: Conduct a pilot test or small-scale implementation of the technology in a controlled environment. This allows for evaluating its performance, identifying potential issues and making necessary adjustments before full-scale implementation. The pilot test should involve representatives from different functional areas and departments to gather feedback and insights.
- Implementation planning: Develop a detailed implementation plan that outlines the timeline, tasks, resource allocation and responsibilities. Define key milestones and set realistic targets. Identify potential risks and develop mitigation strategies. Coordinate with internal teams, external vendors and any required consultants or experts to ensure a smooth implementation process.
- Training and change management: Provide comprehensive training programs to employees to familiarize them with the new technology and its usage. Ensure that employees understand the benefits, processes and potential impact on their roles and responsibilities. Implement change management strategies to address any resistance or concerns among the workforces.
- Deployment and integration: Execute the full-scale implementation of the technology. This may involve hardware installation, software configuration, data migration and integration with existing systems. Test the integration to ensure smooth data flow and interoperability between different systems and departments.
- Performance monitoring and optimization: Continuously monitor the performance and effectiveness of the implemented technology. Track key performance indicators (KPIs) to assess the impact on productivity, efficiency, cost reduction and other desired outcomes. Make necessary adjustments, refinements, or enhancements to optimize the technology's performance.
- Continuous improvement: Foster a culture of continuous improvement by encouraging feedback from employees, customers and other stakeholders. Gather insights and learnings from the implementation process to drive further enhancements and refinements. Stay informed about emerging technologies and industry trends to identify opportunities for future technology adoption and innovation.

By following this procedure, BBT Steel integrated Industry 4.0 technologies into their company, which unlocked operational efficiencies, improved productivity and maintained their competitiveness.

## Key learnings

This case study of BBT Steel Manufacturers has provided relevant information on how to develop an Industry 4.0 framework to integrate business and manufacturing systems. The case provides significant information regarding the process of integrating Industry 4.0 in a steel manufacturing company. The findings from this case study can be used by the identified stakeholders to address similar issues in any manufacturing industry.

REFERENCES

Mousa E, Wang C, Riesbeck J and Larsson M (2016) Biomass applications in iron and steel industry: An overview of challenges and opportunities. *Renewable and Sustainable Energy Reviews* **65**: 1247–1266.

Olalekan O, Afees O and Ayodele S (2016) An empirical analysis of the contribution of mining sector to economic development in Nigeria. *Khazar Journal of Humanities & Social Sciences* **19(1)**.

Pandia O, Sărăcin I, Dobrotă D, Ganea-Christu I and Marin G (2016) The Management of Steel Industry in the context of durable development. *Annals of the Faculty of Engineering Hunedoara* **14(4)**: 179.

Stock T and Seliger G (2016) Opportunities of sustainable manufacturing in industry 4.0. Procedia CIRP **40**: 536–541.

Arnesh Telukdarie
ISBN 978-1-83608-365-8
https://doi.org/10.1108/978-1-83608-362-720241005


Chapter 5

# The future of small to medium enterprises (SMEs) in digital transformation: a case of food and beverage (FoodBev) SMEs in South Africa

A Telukdarie, M Munsamy and T Dube

## Introduction

As far as businesses go, there are always hurdles that need to be overcome. Starting a business itself is a big achievement for entrepreneurs, but the main challenge is maintaining one. The business environment is characterised by a dynamic and constantly changing nature. It is subject to evolution and innovation, leading to the emergence of novel patterns. To sustain growth, small enterprises must remain abreast of evolving trends. However, this is the juncture where several enterprises begin to disintegrate. Timing is crucial for businesses to effectively market their products following emerging trends. This implies that they must continue conducting market research to anticipate their forthcoming products. If a small business is unable to meet demand during the height of a trend, it will incur significant financial losses. This is the everyday struggle for SMEs in the FoodBev manufacturing sector in South Africa. The Food and Beverage Sector Education Training Authority (SETA) has long identified SMEs as effective agents of sustainable economic growth and development. Although this is the case, the FoodBev SETA has also observed these SMEs struggle to keep their businesses afloat. This became more apparent during the global COVID-19 pandemic, which brought about unforeseen changes, such as halting global supply chains, bringing the world to the stage of uncertainties.

Prior to the onset of the COVID-19 pandemic, small businesses in the country were already encountering substantial challenges. The decline of the economy, in conjunction with multiple downgrades in ratings that have persistently impacted SMEs, seeing this happening the FoodBev SETA, the mandated authority for promoting, facilitating and incentivizing skills development in the FoodBev manufacturing sector, resolved to deploy strategic initiatives to support SME growth and development. The FoodBev manufacturing SETA, which had recently appointed a chairperson (chair) for digital transformation from the University of Johannesburg, thus instructed the chair to identify and develop tools to support growth and sustainability of SMEs. The chair adopted a mixed methods approach of training, surveys and a global literature scan to identify an appropriate tool for SMEs. Following a structured approach, digital systems were identified as a key requirement from SMEs, thus the chair developed an SME-specific ERP. As a start, four key functionalities were built into the SME ERP platform invoicing, website development, marketplace and business case developer. These functionalities enable the SME to address challenges such as increased presence of business, tracking of purchases, application of funding and so on.

## SMEs are the backbone of South Africa's economy – not without challenges

SMEs are present in all sectors of the economy and are characterised by their diversity and heterogeneity. These types of businesses are defined by either the number of employees or the annual turnover. According to the sector skills plan of FoodBev SETA (2020), SMEs constitute over 98% of businesses in South Africa, provide employment to 50–60% of the country's workforce across all sectors and contribute to a quarter of job growth in the private sector. FoodBev manufacturing is a significant economic activity in South Africa, contributing 26.6% to the overall manufacturing industry (FoodBev SETA, 2020). SMEs in the FoodBev Manufacturing sector, encounter various challenges including competition from larger businesses, limited financial resources, challenges with remote work, limited access to technology, difficulty obtaining credit/loans/funding, limited skills, policies and legal frameworks, limited access to global markets, and inadequate access to facilities such as space and electricity.

## Opportunities and benefits of fourth industrial revolution (4IR) – why 4IR?

The 4IR offers numerous possibilities to improve manufacturing business performance from financial, process, to environmental perspectives (Malang, 2021). Digitalisation has led to the increased adoption of information technology (IT) and other technological tools as drivers of success across various industries. Hence, SMEs must utilise the opportunities digitalisation presents to sustainably grow their business. Failure to enhance their capabilities may result in the loss of their competitive edge. The challenge of meeting the constantly evolving customers' expectations requires a dynamic strategy.

## Advancing chairperson skills in 4IR: the University of Johannesburg and FoodBev SETA collaboration

In August of 2021, the FoodBev SETA asked its research chair at the University of Johannesburg to find strategies and tools to support SMEs to adapt to the digital world as well as implement and utilise the tools that digitalisation offers. Thus, the FoodBev SETA and the research chair took on the strategy of digital transformation to assist the SMEs.

As such the FoodBev SETA and the University of Johannesburg (UJ) invited all FoodBev SMEs to join the blended learning workshops to identify their current and desired digital status to identify support mechanisms and tools. The FoodBev chair was responsible for the development and delivery of the SME 4IR training and identifying initiatives to support the SME; 220 SMEs participated in this training initiative. To gain the relevant information and ensure interactive training sessions, the training was conducted in cohorts of 20 SMEs.

The workshops included both practical and theory sessions. Further to this, each SME was required to complete a questionnaire on their current and desired digital status. The SMEs were provided with the questionnaire at the start of the workshop and required to complete the respective session, upon completion of a specific training session.

The training focused on six business functional areas: strategy, planning, human resources (HR), finance, maintenance and logistics. As business processes are a key navigator of digitalisation, the workshops also included a session on business processes. The workshops also included a session on the green economy, given the sustainability challenges SMMEs are encountering. A summary of the content is shown in Table 5.1.

## Results from the SME 4IR training

The training was interactive, which enabled the identification of challenges being encountered by SMEs. The challenges defined by the SMEs included

- lack of digital literacy
- lack of understanding of technology-enabled services and products
- time constraints
- loadshedding
- space requirements
- limited finances
- limited resources from skilled personnel to equipment.

For each training session (per functional area), the SME completed a survey, which identified the current and desired (to-be) state of operation – that is, were the activities conducted manually or by way of the use of a system? The survey is attached in Appendix A.

For each functional area, the SME identified how performance monitoring, planning, financial monitoring and document management are conducted, as well as the desired state. A Likert scale was utilised to identify the current and desired stated, as per the rating detailed in Table 5.2. The results from the questionnaire were captured in Microsoft Excel to define the current and future state of the SME. Figure 5.1 defines the results for the status while Figure 5.2 presents the desired status.

### A. Current status

The questionnaire found that the majority of the SMEs did not have any system to monitor performance, conduct planning, track financial transactions and manage documentation; 71% of SMEs did not have a system to monitor performance for operations, 68% for finance, 71% for marketing, 71% for maintenance and 80% for HR. A few SMEs have a fully integrated system, ranging from 0% to 13%, with the focus being on financial activities. See Figure 5.1.

### B. Desired status

The SMEs were then required to identify their desired status. The results, as per Figure 5.2, illustrate scepticism among SMEs on the use of digital tools and systems to improve growth and sustainability; 43% to 46% indicated no desire to digitalise, even when presented with verified benefits of digitalisation (as per training). However, there were SMEs interested in digitalising, with 12% to 24% indicating the desire for at least a basic software-based system, with 12% to 40% indicating the need for a fully integrated system. Again, the highest demand for fully integrated functions was for the finance functionalities. The analysis of the current and desired states identifies the financial function as a critical area of support for SMEs, which is supported by the literature. SMEs in South Africa are financially constrained and struggle from managing financial transactions to record-keeping financial activities to acquiring loans or funds.

## The solution for the SMEs

Based on new findings of the survey the research chair identified 10 systems that would support the SMEs in their digitalisation journey, as well as provide an entry point for the sceptical SMEs, Figure 5.3 shows the architecture of the applications below:

- business case systems
- digital invoicing and cash flow
- digital web creation
- auto marketing
- market and technological scanning
- international marketing
- order management
- digital product identification
- digital technologies
- document management system (DMS).

## Conclusion

The future of SMEs in the FoodBev industry in South Africa lies in their successful adoption of digital transformation. The case study of FoodBev SMEs has shed light on the challenges they face and the immense opportunities presented by embracing digital technologies. The study highlighted that the rising costs of inputs, coupled with the need for technical expertise, pose significant challenges for FoodBev SMEs. However, by implementing innovative solutions, such as investing in training and skill development, SMEs can address these challenges effectively. Furthermore, by leveraging the power of digital transformation, SMEs can overcome challenges, tap into new markets, and pave the way for sustainable growth and competitiveness in the dynamic and digitally driven business environment. Through collective efforts and strategic actions, FoodBev SMEs are poised to thrive and make a lasting impact in the evolving landscape of the FoodBev industry in South Africa.

## Core reading

Autio E, Nambisan S, Thomas LDW and Wright M (2018) Digital affordances, spatial affordances, and the genesis of entrepreneurial ecosystems. *Strategic Entrepreneurship Journal* **12**(**1**, July 2017): 72–95. https://doi.org/10.1002/sej.1266

Foodbev SETA (2020) Sector Skills Plan.

Malang UN (2021) *Entrepreneurship, Trade and Exchange Activities* **2**: 101–110. https://doi.org/10.26661/2522-1566/2021-1/16-08

## Further readings

Ziółkowska MJ (2021) Digital transformation and marketing activities in small and medium-Sized enterprises. *Sustainability* **13(5)**: 2512–2516. https://doi.org/10.3390/su13052512

## Exhibits

Table 5.1 Workshop content (continued on next page)

| Workshop content | |
|---|---|
| Strategy<br>• Understand the importance of business strategy.<br>• The need for strategic analysis.<br>• Importance of digital intervention and digital strategy.<br>• Fundamental steps towards technology adoption. | Business Processes<br>• What is a business process?<br>• Application in businesses. |
| Operations Function<br>• Familiarise with the operations strategies for small businesses.<br>• Understand the process of creating a sustainable plan for the operations strategies.<br>• Describe the implementation process and the major approaches to the operations strategies.<br>• Familiarise with some digitalisation best practices. | Finance Function<br>• Familiarise with the financial strategies for small businesses.<br>• Understand the process of creating a sustainable plan for the finance strategies.<br>• Describe the implementation process and the major approaches to the finance functional strategies.<br>• Familiarise with some digitalisation best practices. |
| Marketing Function<br>• Understand the importance of marketing.<br>• Understand the process of marketing your products and services.<br>• Understand the digital marketing framework and process.<br>• Familiarise with digital marketing solutions.<br>• Co-develop a digital marketing strategy for your company. | Maintenance Function<br>• Understand what maintenance management is.<br>• Understand the different maintenance management strategies.<br>• Understand the maintenance management phases.<br>• How to develop maintenance management objectives.<br>• Co-develop a digital maintenance management strategy for your company. |

Table 5.1 Continued

| Workshop content | |
|---|---|
| Human Resource Function<br>• Define HR management.<br>• HR management processes.<br>• Familiarise with digital HR solutions. | Green Economy<br>• What is the green economy? |

Table 5.2 Likert scale

| | |
|---|---|
| S1 | No-system |
| S2 | Paper system: no continuous assessments |
| S3 | Paper system: continuous assessments |
| S4 | Basic software-based system |
| S5 | Fully integrated system |

Figure 5.1 Current status of the system per function (author's own)

| | Which type of system do you use to monitor performance in each function? | | | | | What type of system do you use to conduct planning? | | | | | What type of system do you use to track financial transactions? | | | | | What type of system do you use for document management? | | | | |
|---|---|---|---|---|---|---|---|---|---|---|---|---|---|---|---|---|---|---|---|---|
| | Operations | Finance | Marketing | Maintenance | HR | Operations | Finance | Marketing | Maintenance | HR | Operations | Finance | Marketing | Maintenance | HR | Operations | Finance | Marketing | Maintenance | HR |
| S1 | 71% | 68% | 71% | 71% | 80% | 58% | 55% | 59% | 61% | 67% | 57% | 56% | 61% | 64% | 61% | 52% | 55% | 59% | 62% | 55% |
| S2 | 0% | 0% | 0% | 1% | 0% | 22% | 13% | 13% | 13% | 13% | 0% | 1% | 1% | 0% | 3% | 9% | 8% | 8% | 8% | 11% |
| S3 | 22% | 17% | 19% | 17% | 9% | 0% | 9% | 9% | 9% | 4% | 15% | 15% | 13% | 13% | 14% | 21% | 22% | 17% | 17% | 17% |
| S4 | 7% | 9% | 5% | 5% | 5% | 20% | 20% | 17% | 17% | 16% | 16% | 15% | 15% | 13% | 13% | 13% | 9% | 9% | 6% | 13% |
| S5 | 0% | 7% | 4% | 7% | 7% | 0% | 3% | 3% | 0% | 0% | 13% | 13% | 10% | 10% | 10% | 6% | 6% | 7% | 7% | 3% |

Figure 5.2 Desired status of the system per function (author's own)

| | What level of system maturity would maximise performance In? | | | | | What level of system maturity would maximise planning In? | | | | | What level of system maturity would maximise document management? | | | | | What level of system maturity would maximise business financial value? | | | | |
|---|---|---|---|---|---|---|---|---|---|---|---|---|---|---|---|---|---|---|---|---|
| | Operations | Finance | Marketing | Maintenance | HR | Operations | Finance | Marketing | Maintenance | HR | Operations | Finance | Marketing | Maintenance | HR | Operations | Finance | Marketing | Maintenance | HR |
| S1 | 46% | 46% | 46% | 46% | 46% | 51% | 51% | 51% | 51% | 51% | 45% | 45% | 45% | 45% | 45% | 43% | 43% | 43% | 43% | 43% |
| S2 | 5% | 5% | 5% | 5% | 15% | 0% | 0% | 0% | 0% | 10% | 9% | 9% | 9% | 9% | 19% | 0% | 0% | 0% | 0% | 10% |
| S3 | 1% | 1% | 1% | 11% | 1% | 9% | 9% | 9% | 17% | 7% | 0% | 0% | 0% | 13% | 0% | 5% | 5% | 5% | 15% | 13% |
| S4 | 17% | 17% | 22% | 27% | 23% | 11% | 11% | 17% | 16% | 16% | 12% | 12% | 21% | 12% | 15% | 12% | 12% | 21% | 15% | 15% |
| S5 | 31% | 32% | 27% | 12% | 16% | 29% | 29% | 23% | 16% | 16% | 34% | 34% | 25% | 21% | 21% | 40% | 40% | 31% | 27% | 19% |

Figure 5.3 Architecture of the applications developed (author's own)

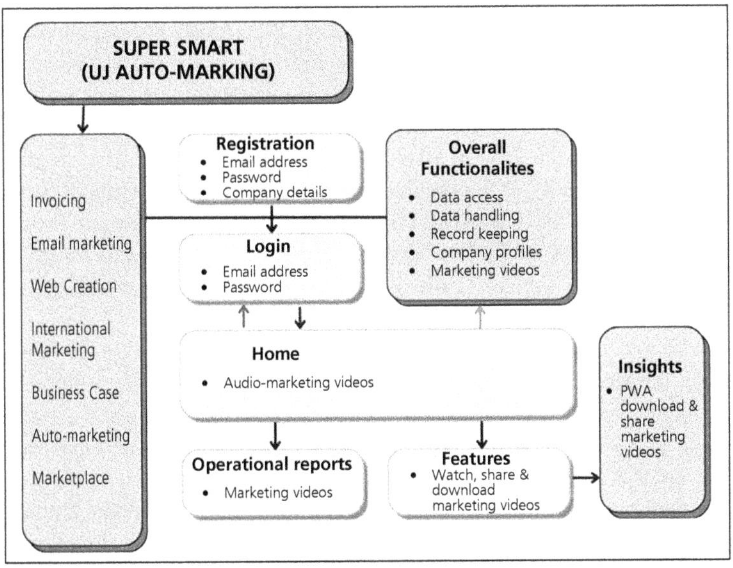


Figure 5.4 Snapshots of the applications. Courtesy of University of Johannesburg

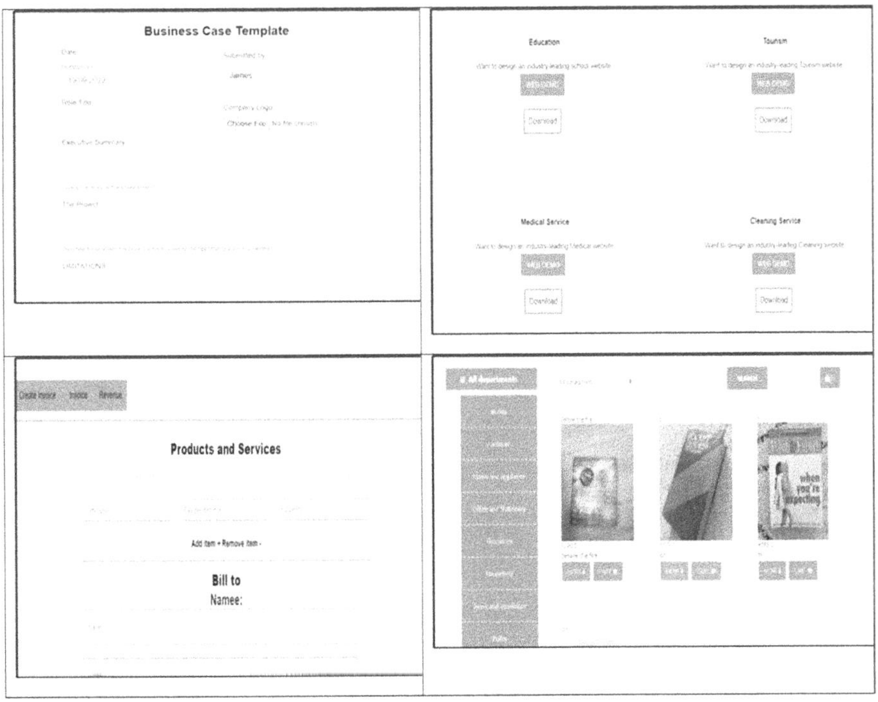

# Teaching note

## Synopsis

The SME case study is based on the SMEs in the South African FoodBev industry. These SMEs typically comprise about five to ten employees. The FoodBev SETA acknowledges these SMEs as crucial in contributing to employment, entrepreneurship as well and industrial development. The SETA has noted that most companies that adopt digital technologies perform better than those that do not (Autio *et al.*, 2018), and believes that digital technologies boost the position of SMEs by broadening the scope of their activities and fostering innovation. The SETA has identified that although these companies are eager to adopt digital systems/technologies, they often face challenges, including finance, skills, resources and technical proficiency. Another challenge is the limited training availability for SMEs concerning digital technologies.

After analysing the situation, FoodBev SETA came up with a solution to the problem that SMEs face. The first step that FoodBev took was to partner with the University of Johannesburg, where the SETA appointed a research chair from the University of Johannesburg. The SETA and the research chair then compiled a training package for SMEs, with a special focus on the 4IR technologies/digital systems that have the potential to contribute to the growth of SMEs. While the training sessions provided knowledge to SMEs about digital systems, the University of Johannesburg was able to employ a tool that measured the current state of the SMEs with digital systems. This enabled the University of Johannesburg to be able to identify which systems were mostly required by the SMEs. Therefore, the University of Johannesburg developed the following systems, see Figure 5.4:

- invoicing
- website builder (DIY)
- business case
- auto-marketing
- email marketing
- international marketing
- marketplace.

## Teaching objectives

The case is written for use in an entrepreneurship module at the undergraduate and postgraduate levels. The focus of the case is to inform the students of the challenges encountered by SMEs and to identify possible solutions.

- Students should be able to illustrate the importance of SMEs in the economy.
- Students should be able to define the challenges that SMEs encounter.
- Students should be able to find solutions to the challenges encountered by the SMEs.

## Research methodology

This case study focuses on the development of training content for SMEs in the context of 4IR. The case adopts a mixed method approach, combining secondary data from academic publications with survey data gathered during 4IR training by the University of Johannesburg.

The purpose of this study is to enhance the digital awareness and overall system maturity of SMEs to adapt to the rapidly changing business landscape. The study utilised survey data collected during the 4IR training conducted by the University of Johannesburg. Out of the 200 companies trained, 150 surveys were received and considered for analysis. The surveys were designed to capture the opinions of SME participants regarding their current system maturity and their envisaged required system maturity. The surveys employed a rating scale defining five sequential maturity stages (S1 to S5) to assess both current and required system maturity, as shown in Table 5.3 below.

Table 5.3 System rating scale

| | |
|---|---|
| S1 | No-system |
| S2 | Paper system: no continuous assessments |
| S3 | Paper system: continuous assessments |
| S4 | Basic software-based system |
| S5 | Fully integrated system |

## ERP system development

After analysing the questionnaire survey data, the researchers identified several digital systems crucial for SME operations, including invoicing, website builder, business case, auto-marketing, email marketing, international marketing and market place. The goal was to create an integrated ERP system to streamline and enhance the efficiency of these operations. Before integrating the platforms into an ERP system, each digital system was developed as an independent entity. This step allowed the development team to ensure that each platform was fully functional and optimised for its specific purpose. The front-end technology stack was chosen consistently across all platforms to maintain uniformity and ease of development. CSS, HTML and JavaScript were selected as the primary front-end technologies. These languages are widely used, and the development team possesses the necessary skills to work efficiently with them. For the back-end development, the researchers opted for the Python-based Django full-stack framework. Django was chosen due to its comprehensive nature and rich set of libraries, which expedited the development process. Additionally, the team's familiarity with Python further contributed to the decision. One of the significant advantages of Django is its ability to facilitate seamless integration of diverse platforms. This feature was crucial for creating a cohesive ERP system that could effectively communicate and share data across all modules. The researchers selected the SQL Server as the database to store all data sets for the ERP system. SQL Server is a widely utilised and robust database server, known for its reliability and performance. Additionally, it offers the advantage of being available as an open-source solution, which is aligned with the project's budget constraints.

## Keywords

SMEs, digitalisation, digital technologies, ERP systems, business functions, 4IR and Digital transformation.

## Assignment questions

1. What is the contribution of SMEs in the FoodBev SETA to the economy?
2. Critically analyse the challenges faced by SMEs in the FoodBev industry by way of a PESTEL analysis.
3. What are the opportunities that digital systems present for SMEs?
4. Analyse the results of the survey and identify the key findings.
5. Identify the average percentage of SMEs that require a fully integrated system in maximising performance and conducting planning. (Calculate the average.)
6. In your own words elaborate on what each module of the SME ERP did.
7. What solutions do you propose to resolve any two of the challenges identified by the SMEs? (Open question.)
8. Identify the software used in developing the SME ERP.
9. Do you think the ERP system developed is specific to the SMEs in the FoodBev industry or can it be used by an SME? Justify your answer.

## Additional materials

The following references are recommended for more information.

Akileswaran K and Hutchinson G (2019) *Adapting to the 4IR: Africa's Development in the Age of Automation*. Tony Blair Institute for Global Change. England and Wales, 50.

Autio E, Nambisan S, Thomas LDW and Wright M (2018) Digital affordances, spatial affordances, and the genesis of entrepreneurial ecosystems. *Strategic Entrepreneurship Journal* (no. July 2017): 72–95. https://doi.org/10.1002/sej.1266

Foodbev SETA (2020) *Sector Skills Plan*. Food and Beverages Manufacturing SETA.

Ingley C, Khlif W and Karoui L (2017) SME growth trajectories, transitions, and board role portfolios: a critical review and integrative model. *International Small Business Journal: Researching Entrepreneurship* **35(6)**: 729–750. https://doi.org/10.1177/0266242616680280

Malang UN (2021) *Entrepreneurship, Trade and Exchange Activities* **2**: 101–110. https://doi.org/10.26661/2522-1566/2021-1/16-08

Mokonyama, M, Malatji M and Mlitwa N (2022) Addressing operational challenges of small and medium enterprises of the logistics industry-potential for autonomous vehicles. In *8th International Conference on Engineering and Emerging Technologies, ICEET 2022*. University of South Africa (Unisa), Graduate School of Business Leadership (SBL), Midrand, South Africa. https://doi.org/10.1109/ICEET56468.2022.10007250

Wangmo C and Islam SMN (2021) *SME Finance: Constraints and an Information Asymmetric Perspective. SME Finance: Constraints and an Information Asymmetric Perspective*. Bhutan Care Credit Ltd., Microfinance Project, His Majesty's Secretariat, Thimphu, Bhutan. https://doi.org/10.52305/YNJH5087

Ziółkowska MJ (2021) Digital transformation and marketing activities in small and medium-sized enterprises. *Sustainability (Switzerland)* **13(5)**: 2512. https://doi.org/10.3390/su13052512

## Teaching plan and timing

This teaching plan aims to educate participants about the significance of digital transformation for small and medium-sized enterprises (SMEs) in the FoodBev industry in South Africa.

The course will explore the challenges faced by FoodBev SMEs, the opportunities digital transformation presents and strategies to drive successful digital adoption. The teaching plan is designed to be delivered over two days, with each day comprising four sessions, each lasting for 1.5 hours.

## Day 1

### SESSION 1: INTRODUCTION AND OVERVIEW

- Introduce the topic of digital transformation and its relevance to SMEs in the FoodBev industry.
- Present key trends and statistics related to digital transformation in South Africa.
- Highlight the challenges faced by FoodBev SMEs and the potential impact of digitalisation.

### SESSION 2: CHALLENGES AND OPPORTUNITIES

- Conduct a detailed PESTEL analysis to identify the external factors influencing FoodBev SMEs in South Africa.
- Discuss the specific challenges posed by each factor and how they can be addressed through digital transformation.
- Explore the opportunities that digital systems present to overcome these challenges and drive growth.

### SESSION 3: CASE STUDY ANALYSIS

- Present the case study of FoodBev SMEs in South Africa, showcasing real-world examples of SMEs embracing digital transformation.
- Analyse the digital strategies employed by successful SMEs and the impact on their business growth and competitiveness.
- Encourage group discussions and interactive exercises to engage participants in the case study analysis.

### SESSION 4 (DIGITAL TRANSFORMATION STRATEGIES):

- Introduce participants to various digital transformation strategies applicable to FoodBev SMEs.
- Discuss the importance of customer-centric approaches, data analytics, e-commerce and online marketing.
- Guide participants in developing their digital transformation roadmaps based on their specific business contexts.

## Day 2

### SESSION 5: SKILL DEVELOPMENT AND WORKFORCE EMPOWERMENT

- Emphasize the importance of upskilling the workforce to effectively leverage digital technologies.
- Discuss various training and development programs available for SME employees.
- Share best practices on how to create a culture of innovation and continuous learning within SMEs.

#### SESSION 6: COLLABORATION AND PARTNERSHIPS

- Explore the potential benefits of collaboration and partnerships with technology providers, digital platforms and other SMEs.
- Discuss success stories of SMEs that have achieved growth through collaboration.
- Facilitate group exercises to encourage participants to identify potential collaboration opportunities.

#### SESSION 7: GOVERNMENT SUPPORT AND POLICY FRAMEWORKS

- Discuss the role of government in supporting digital transformation for SMEs in South Africa.
- Highlight relevant policies, incentives and funding opportunities available for SMEs.
- Guide on accessing government support programs.

#### SESSION 8: CONCLUSION AND ANSWERING QUESTIONS OF THE CASE STUDY

- Summarise the key learnings from the course.
- Facilitate a group discussion to develop individualised action plans for participants to implement digital transformation strategies in their respective FoodBev SMEs.
- Encourage participants to share their action plans and commitments to drive digitalisation in their businesses.
- Provide the questions for the students to answer.

## Analysis of the questions

1. It is important to note that the specific contribution of SMEs in the FoodBev SETA to the economy may vary based on the size, scale and geographic distribution of these businesses.
2. A PESTEL analysis is a framework used to analyse the external macro-environmental factors that can impact a business or industry. Let's apply the PESTEL analysis to critically analyse the challenges faced by SMEs in the FoodBev industry.
    - Political factors: SMEs in the FoodBev industry often face complex and stringent regulations regarding food safety, labelling, packaging and health standards. Compliance with these regulations can be burdensome and costly for smaller businesses. Changes in trade policies, tariffs and import/export regulations can impact the supply chain and sourcing of ingredients for FoodBev SMEs, leading to fluctuations in production costs and pricing.
    - Economic factors: SMEs are more vulnerable to economic fluctuations, including recessions and inflation. A downturn in the economy can lead to reduced consumer spending, affecting the demand for FoodBev and potentially reducing sales for SMEs in the industry. Rising costs of raw materials, ingredients and transportation can squeeze profit margins for SMEs in the FoodBev sector, making it challenging to maintain competitiveness.
    - Social factors: Consumers' preferences for healthier, organic and sustainable food products are evolving rapidly. SMEs in the FoodBev industry must adapt their product offerings and marketing strategies to meet these changing demands.

Increased awareness of health issues, such as obesity and food allergies, can lead to more stringent labelling requirements and consumer demand for healthier food options, putting pressure on SMEs to reformulate their products.

- Technological factors: Technological advancements in food processing, packaging and distribution may require significant investments for SMEs to adopt automation, leading to potential challenges in terms of resource availability and technical expertise. SMEs need to establish a robust online presence and e-commerce capabilities to reach a broader customer base. However, digital transformation can be challenging for smaller businesses with limited resources.
- Environmental factors: Increasing awareness of environmental issues requires SMEs in the FoodBev industry to adopt sustainable practices, reduce food waste and implement eco-friendly packaging solutions, which may involve additional costs. Climate-related events, such as extreme weather conditions and natural disasters, can disrupt the supply chain, affecting the availability and cost of raw materials for SMEs.
- Legal factors: SMEs in the FoodBev industry must safeguard their recipes, brand names and product designs from imitation and infringement, necessitating legal efforts and resources. Compliance with labour laws, minimum wage requirements and employee benefits can pose challenges for SMEs in managing their workforce effectively while controlling operational costs.

3. Digital systems present a wide range of opportunities for SMEs to transform their businesses, enhance operations and compete effectively in a rapidly evolving digital landscape. By leveraging these opportunities, SMEs can strengthen their market position, drive growth and unlock new avenues for success.
4. The analysis of the current and desired states identifies the financial function as a critical area of support for SMEs, which is supported by the literature. SMEs in South Africa are financially constrained and struggle from managing financial transactions to record-keeping financial activities to acquiring loans or funds.
5. Add all the percentages of S5 divide by 500 and multiply by 100.
   a. 23.6%
   b. 22.6%
6. The Invoice module focuses on managing invoicing and payments, the DIY website module empowers SMEs to create and manage their websites, and the Business Case module supports decision-making and planning for new business projects. Each of these modules enhances the efficiency and capabilities of SMEs, enabling them to operate more effectively and competitively in their respective industries.
7. Open question.
8. The SME ERP system was developed using MySQL and Python to ensure reliability, scalability and security.
9. Open question. Possible answer: While the initial design of the ERP system might have considered the specific needs of SMEs in the FoodBev industry, the system's architecture and core functionalities suggest that it can be effectively used and customised for SMEs in other industries as well. The flexibility and adaptability of the ERP system make it suitable for various business domains, providing SMEs with the opportunity to leverage its benefits regardless of their industry.

## Key learnings

The case study on FoodBev SMEs in South Africa demonstrates that digital transformation is a fundamental aspect of their future growth and success. Embracing digital technologies, customer-centric approaches and innovation will empower SMEs to navigate challenges, capitalise on opportunities and stay competitive in a digitally driven business environment. The study reveals that digital transformation offers significant potential for growth and improved efficiency for FoodBev SMEs. By adopting digital systems and processes, SMEs can streamline operations, enhance productivity and access new markets, driving business expansion. The case study highlights the significance of providing adequate training and skill development to employees in FoodBev SMEs to effectively leverage digital technologies. Upskilling the workforce is essential for maximising the benefits of digital transformation. Furthermore, the case study highlights the significance of providing adequate training and skill development to employees in FoodBev SMEs to effectively leverage digital technologies. Upskilling the workforce is essential for maximising the benefits of digital transformation. Finally, the case study indicates the potential for collaboration and partnerships with technology providers, digital platforms and other SMEs. Collaborative efforts can pool resources and expertise, leading to mutual benefits and shared growth opportunities.

REFERENCES

Autio E, Nambisan S, Thomas LDW and Wright M (2018) Digital affordances, spatial affordances, and the genesis of entrepreneurial ecosystems. *Strategic Entrepreneurship Journal* (no. July 2017): 72–95. https://doi.org/10.1002/sej.1266

Foodbev SETA (2020) *Sector Skills Plan*. Food and Beverages Manufacturing SETA.

Malang UN (2021) *Entrepreneurship, Trade and Exchange Activities* **2**: 101–110. https://doi.org/10.26661/2522-1566/2021-1/16-08

emerald PUBLISHING
ice Publishing

Arnesh Telukdarie
ISBN 978-1-83608-365-8
https://doi.org/10.1108/978-1-83608-362-720241007


# Chapter 6
# Digitalisation of data gathering for childhood development in Africa

**A Telukdarie, R Mongwe and LPL Graham**

## South African Research Chairs Initiative (SARChI)

In 2006, the South African government embarked on a noble mission to attract and retain research talent in South Africa and build research capacity in areas of national and strategic importance. They launched the South African Research Chairs Initiative (SARChI), a visionary project that supports studies in a variety of disciplines, including the humanities, social sciences, engineering and natural science. Outstanding researchers with a remarkable history of research excellence and the potential to lead their fields internationally are appointed as SARChI chairpersons. Typically, these eminent scholars are appointed for a five-year term with the option of extension for an additional five years.

One of the first SARChI chairpersons is the esteemed Professor Leila Patelin, a leading figure in the field of social development and welfare. Professor Patelin holds the South African Research Chair in Welfare and Social Development and is also the founder and director of the Center for Social Development (CSDA) at the University of Johannesburg. Through her appointment as a SARChI chair, Professor Patelin is dedicated to enhancing the social well-being of Africans, especially children.

In the field of engineering, Professor Arnesh Telukdarie from UJ brings their innovative spirit to the forefront. Their chair allows them to develop digital solutions to mitigate various challenges. They previously held a senior managerial position at Sasol, one of the most innovative multinational companies in Africa, and are currently a professor of digital business at the Johannesburg Business School.

SARChI also recognises education and care in childhood as an area of national and strategic importance. Professor Jace Pillay received a chair in education and care to continue the mission of improving the educational, psychological and social care of orphans and vulnerable children.

Moreover, Professor Elizabeth Henning chairs the integrated studies of learning language, mathematics and science in primary school. They have been a leading researcher in mathematical cognition and language in the foundation and lower primary phases of school. Additionally, they focus on primary school children's learning of science concepts and reading science textbooks, combining instructional film production with research to improve learning outcomes.

Professor Shane Norris, a global health professor at the University of Southampton and a research professor at the University of Witwatersrand in the Department of Pediatrics, focuses on developing interventions that improve maternal and child health outcomes.

Over the years, the SARChI initiative has flourished, creating an interconnected community of scholars who share knowledge, collaborate and inspire each other. An example of this is the collaboration between Professor Leila Patel, Jace Pillay, Elizabeth Henning, Arnesh Telukdarie and Shane Norris.

This group of passionate and determined chairpersons embarked on a mission that would change the lives of children across the country. They faced a significant challenge in understanding and addressing the obstacles that these children faced daily. After careful consideration, the chairpersons unanimously agreed that monitoring children's well-being in areas such as health, nutrition, education, safety and security, and psycho-social well-being was crucial to tackling these challenges.

Their vision was clear, but they quickly realised that traditional paper-based data collection methods would not do justice to their noble cause. The drawbacks of poor data quality, redundancy and wasted time were far too great. The ambitious goal of gathering information from 163 children across these vital domains required a modern solution. The chairs decided to embrace the wonders of digital data collection. They understood that it might demand a greater investment of time and resources initially, but the long-term benefits far outweighed the risks. With the support of engineering expertise, they were confident they could create a robust digital platform that would streamline the entire process and ensure the data's accuracy and efficiency. Although their hearts were filled with both excitement and concern, the financial implications of this revolutionary approach loomed large. Still, they knew that they could not let such obstacles deter them from their mission. They were driven by a collective commitment to making a difference in the lives of these vulnerable children.

With unwavering determination, the chairpersons formed a harmonious community of practice, bringing together a diverse array of professionals from various disciplines. From social work, sociology and psychology to education experts, mathematicians and language curriculum specialists, each person played a unique and crucial role in this interdisciplinary endeavor.

## The launching of the Community of Practice

The transformative initiative took root in South Africa with generous funding from the National Research Foundation. The Community of Practice (CoP) model was adopted in 2018 to encourage more integrated and synergistic interventions to improve children's well-being. Community of Practice (Vincent *et al.*, 2018; Wenger *et al.*, 2002) are social organisations comprising like-minded individuals who share a concern, a set of problems or a passion for a topic, and who deepen their knowledge and expertise in this area through regular interaction. Additionally, they innovate and advance a particular field. They function as 'social learning systems where practitioners connect to solve problems, share ideas, establish standards, develop tools, and build relationships with peers and stakeholders' (Snyder *et al.*, 2004). The professors strongly believe that the CoP model will help reduce the overall cost of this initiative since all the skills are housed under one roof. This CoP brought together a community from multi and transdisciplines. This CoP, through their collaborative efforts, developed a digital tracking tool as well as an intervention plan for child well-being.

## Tracking the well-being of children in South Africa

The professors believe that monitoring children's well-being can help identify areas where they may be struggling, such as access to education, health care and nutrition. This information can then be used to prioritise resources and interventions to address their needs. Tracking the well-being of children can also help evaluate the effectiveness of policies and programs designed to support them. Policymakers and program implementers can determine whether their interventions are having a positive impact and make necessary adjustments by collecting data on key indicators. Data on the well-being of children can be used to inform advocacy efforts aimed at improving the lives of children in South Africa and to raise awareness about the issues facing them.

## Data collection without a digital platform

Based on previous experiences, the professors understand that paper-based data collection is often inexpensive compared to alternative methods, such as online surveys, which require more resources to set up. Paper surveys are easy to understand and complete, making it simple for people of all ages to participate. Additionally, with paper surveys, participants can complete the survey at any time and place, making it easier to reach a wider audience. Furthermore, paper surveys generate a tangible record of responses that can be stored and analysed over time. However, paper-based data collection can be time-consuming and result in errors. Paper surveys have a limited reach and can have an environmental impact due to the need for paper production and disposal. Consequently, the professors are exploring digital tools to potentially minimise costs by insourcing all the technical aspects.

## Digital tools for improving child well-being outcomes

Globally, there is evidence of effective programs and policies that benefit children's development. The 2015 United Nations Sustainable Development Goals (SDGs) add to this ambition, particularly SDG3 (health) and SDG4 (education), both of which are inextricably and fundamentally linked to young children. SDGs mark a global shift from improving child survival rates to improving children's health and developmental trajectories (Goldfeld and Yousafzai, 2018). Considerable inequalities exist globally in access to and quality of services to help young children thrive despite the increasing developmental burden – for example, of the 13 m surviving preterm infants globally, approximately 2.7% have moderate to severe neurodevelopmental impairments and 4.7% have mild neurodevelopmental impairments with poor access to developmental monitoring and early interventions, particularly in resource-constrained settings (Goldfeld and Yousafzai, 2018). SDG4.2 highlights the importance of assessing both access and quality of early childhood care, development and education services for all children. This target has generated high demand for tools that monitor children's development (Goldfeld and Yousafzai, 2018). These tools hold promise in three areas that are relevant to young children and, consequently, SDG3 and SDG4 globally.

- Clinical tools to stimulate greater attention (through early detection and intervention) for children with developmental risks, delays and difficulties.
- Population monitoring of early childhood development with a focus on subpopulations to highlight inequities within and across countries.
- Accelerating the spread and implementation of effective health and education interventions through large-scale evaluations.

Currently, available knowledge on early childhood development is limited, especially in low- and middle-income countries (LMICs). One of the barriers to assessing early childhood development is the scarcity of cross-culturally appropriate tools that are feasible, affordable and have good psychometric properties (Goldfeld and Yousafzai, 2018). Most of the currently available tools were developed in high-income settings, particularly in the USA (such as the Ages and Stages Questionnaire and the Bayley Scales of Infant and Toddler Development) (Goldfeld and Yousafzai, 2018). However, tools used in LMICs, such as the Malawi Developmental Assessment Tool, Rapid Neurodevelopmental Assessment, Kilifi Developmental Inventory and Caregiver Reported Early Development Index, can help to address this important gap (Goldfeld and Yousafzai, 2018). Nevertheless, these tools might not be suitable for a South African setting. Therefore, it is essential to develop and test a tool that is based on a South African setting.

The community of practice (CoP) achieved remarkable milestones by working tirelessly in unity and collaboration. They harnessed the power of their diverse insights to develop a cutting-edge digital tracking tool for monitoring children's well-being. They believed that the tool would be able to accommodate all countries, especially LMICs.

## Development of the child well-being tracking tool (CWTT)

At the University of Johannesburg, a team of specialists and skilled engineers led by Prof. Telukdarie embarked on an exciting project that would revolutionise the way child well-being data was collected. Their goal was to create a digital tool that would streamline surveys, make them accessible to all and, ultimately, improve the lives of children across South Africa. Countless technical conversations echoed through CoP as the team discussed and refined the high-level functional requirements for their innovative creation, the CWTT. To make a real impact, the platform had to be user-friendly, ensuring that even those with limited technological experience could easily navigate and participate in the surveys.

With a passion for open access and widespread adoption, they decided to make the CWTT an open-source platform. This way, anyone who wished to contribute to the cause could do so without barriers, fostering a collaborative effort to uplift children's well-being.

Understanding the diverse roles and perspectives that influence a child's life, they made sure the platform could accommodate data from various role players, including parents, teachers, caregivers, health care professionals and more. By capturing information from multiple sources, they aimed to gain a comprehensive understanding of a child's needs and create more effective interventions.

Privacy and data protection were paramount in their minds, so they incorporated features to ensure compliance with the Protection of Personal Information Act (POPIA). All personal identifiers were securely hidden from open data extractions, safeguarding the children's sensitive information. The team recognised the challenges faced by children living in informal housing and ingeniously included GPS capabilities in the CWTT. This would enable field workers to pinpoint a child's location accurately, allowing targeted interventions and support.

To further enhance data collection, they empowered users to capture images, such as IDs or facial scans of the child. This visual information would add depth and context to the data, facilitating a more comprehensive understanding of each child's unique situation. With a focus on flexibility and adaptability, they ensured that users could define and build questions as needed for any survey instrument. This autonomy eliminated dependency on developers, enabling the CWTT to evolve and cater to specific research requirements.

Seamless integration and linkage of instruments were key to their vision. They developed a brilliant system where multiple expert-linked instruments in a particular study could be sequentially linked together through the child's ID. This interconnectedness streamlined the process and eliminated redundancy. To maintain data accuracy, they included features to validate and edit data during the gathering process. This real-time editing ensured that the information collected was reliable and consistent.

In their quest to empower field workers and researchers, they made the CWTT capable of functioning offline, allowing surveys to be conducted without a data connection. This feature proved vital, especially in remote or underserved areas. The engineering team devised a three-tier design comprising a front-end for questionnaires, a powerful back-end SQL database and a user-friendly interface powered by PowerApps. This architecture utilised existing resources, minimising the need for outsourcing and ensuring cost-effectiveness. The architecture is defined in Figure 6.1.

Users can easily set up surveys and monitor data in real-time using customised Microsoft Excel. The PowerApps platform provides functionality and logic while also rendering the user interface. The software was developed with functional requirements in mind, and survey data is stored on a Microsoft SQL server. Authorised users can retrieve data from the database.

The questionnaire/survey sheet contains a list of users/logins for which the researcher wishes to enable the survey, which, if chosen, makes the survey accessible from mobile devices on which these usernames are logged in. The survey is pushed from the Excel spreadsheet and stored on a central database server once it has been set up.

- The survey questionnaire is downloaded to an authorised mobile device (with the PowerApps container app installed) that logs in to PowerApps. Now the respondent can capture a single-response (data point) survey. First, on each survey, the data capture will collect the respondent's personal information, such picture/picture of ID.
- Name and surname.
- ID number.
- The GPS coordinates are automatically derived from the mobile device.

The application can accept input from an Excel spreadsheet to create a research survey that can accommodate four types of questions/responses.

- Free text or number responses (e.g. '12 345' or '226 Smith Street, Braamfontein').
- Multiple selections (tick-box or list-type questions).
- Single selections (yes/no, or radio button/Likert type questions).
- Date picker (select a specific date, e.g. 2019-12-31).

While completing each survey questionnaire, the mobile device does not need to be connected to the internet; however, while submitting the survey, the mobile device must be online to synchronise the data collected. Now, the data capture process advances to the next data point and collects the same information from the questionnaire for the next respondent. After all respondents have been surveyed and the information has been saved to a cloud database, the researcher can use another button on the master Excel spreadsheet to download the survey results. This dumps the results into the spreadsheet for further analysis and processing.

## Child well-being tracking tool user interfaces

The CWTT is designed to accommodate most user requirements and specifications; for instance, the survey setup is prepared on an Excel spreadsheet to make the CWTT user-friendly. After the survey is set up using the Excel spreadsheet, the respondent logs in using a different interface to complete the survey. This means that the app has two front ends: one for the researcher or surveyor and one for the user or respondent.

As mentioned previously, the survey is conducted on the mobile device when the user launches the PowerApps container app on the mobile device. The login page is the user's initial point of contact with the user interface (UI), and it features standard login fields like username and password. Figure 6.2(a) illustrates the UI of the user login page.

The landing page contains a button for submitting information. It is worth noting that the surveyor uses the Excel front end to generate a username and password for the user. Once logged in, the user is taken to a new screen where they are asked to enter their name, last name, ID number, photo and location. The UI of the personal details is illustrated in Figure 6.2(b). Kyle's input can be seen in the UI. Following the completion and submission of the user's profile information, they will be taken to the survey's opening question. As we've established, the CWTT supports four distinct varieties of inquiry.

Figure 6.3(a) depicts the original form of the drop-down menu used for a single-choice question. Users can select an answer to the question from a list of possible choices presented in a drop-down menu. This is done to guarantee the accuracy of the data and facilitate the survey's flow. There is no upper bound on the potential variety of options that could be presented to the user. The drop-down menu gives the surveyor the freedom to create and upload relevant options (answers) to the questions. The diagram reveals that the drop-down menu supports multiple data type options. Figure 6.3(a) depicts the UI of the drop-down menu for a Yes/No question; Figure 6.3(b) depicts the user interface of the drop-down menu for a question with multiple numerical options; and Figure 6.3(c) depicts the UI of the drop-down menu for a Likert-type question.

Figure 6.4(a) depicts the user interface of the date picker. With the date picker, the user can only enter a date by selecting it from the calendar; any other inputs are disallowed, so the date entered will always appear correctly formatted in the response data. Figure 6.4(b) illustrates the UI for the open-ended question; this input enables users to type free text for open-ended questions. This type of input allows for word or sentence responses, addresses, or even conversation-style text responses. Finally, Figure 6.4(c) illustrates the UI for the multiple-answer selection questions, where users can select multiple answers. One or more of the options in the list may be selected. This will result in a multiline or comma-separated list in the results of the survey, but it limits the options only to possible selections and no free text.

## Survey setup

The software is user-friendly because its front end has a typical Microsoft Excel UI with cells and submit buttons. The Microsoft Excel user interface is password-protected, allowing the engineering team to grant relevant administrators administrator privileges. Administrators with administrative privileges can now construct and develop survey questionnaires without assistance from the development team, as well as create fieldwork credentials. Once trained, field employees can begin data collection. The software can accommodate various role-player data – for example, in the first study conducted using the CWTT, the insights and perspectives of caregivers, teachers, health care professionals and parents regarding the children's well-being were accommodated, and all the data input about individual subjects (children) from all the role-players was interlinked using a single master key (children ID or generated unique identifier). Figure 6.6 depicts the UI utilised by the researchers.

## Linking data

As stated in the previous section, once the survey is complete, the data is stored on a Microsoft SQL database. The database contains multiple tables, but only four are pertinent to this discussion. The first is the survey table. According to database normalisation standards, the table is in the third normal form and has two columns: Survey_ID, Survey_Creator, Survey_Created_Date, Survey_Status (e.g. active or not active) and Survey_Name, with the Survey ID serving as the primary key. The primary key identifies the survey name; this is essential for linking this table to other tables in the database in the future. The second table is the question table. Each survey contains many questions; therefore, this table connects the question to the survey. Following the second normal form of database normalisation standards, the table has Question_ID as the primary key, Survey_ID as the foreign key and Question_Text as a non-key column. The following table is the respondents' table, which contains all survey respondents. The respondent's name, surname, ID number, GPS location and photo were all recorded and saved as non-key columns in this table. This table's primary key is a GUID (globally unique identifier) that is automatically generated and used to identify all respondents. The final table is the responses table, which contains each respondent's response. A snapshot of the table is shown in Figures 6.5–6.7; the Response_ID is the primary key connecting the response to the survey, question and respondent. The last three columns are all foreign keys from the three tables; they are used to connect the data in the other tables to this table (the responses table).

All presented tables have been normalised based on relational database standards. Normalisation of databases is essential because it reduces redundancy and eliminates unwanted insertions, deletions and modifications. Nonetheless, it is challenging to conclude from the data presented in this format. To address this issue, the engineering team merged the data from the four separate tables using SQL query statements and presented the data in a single standard table that is easy to read and comprehend. The data is easier to assess in this format; however, to obtain this representation of the data, one would have to learn basic SQL. Therefore, the team employs the SQL query commands from this section.

## Graphical user interface

Upon analysis, the data recorded in the database should educate social well-being practitioners and other professionals involved in the child's overall social development about the child's current social condition and trigger appropriate intervention if necessary. The preceding subsection demonstrated how data from several tables was consolidated and presented in a more comprehensible format. However, the consolidation process requires some knowledge of SQL. To make the data accessible to relevant personnel without any knowledge of SQL, the engineering team developed a graphical user interface (GUI).

Figures 6.5–6.7 is a screenshot of the graphical user interface; the current view in Figures 6.5–6.7 is equivalent to that shown in Figures 6.5 and 6.6, which depicts the consolidated data from the Microsoft SQL database. The difference in this instance is that the user does not need to execute any SQL statements to retrieve the data; instead, the data can be accessed with a simple button click. The GUI has a search feature, which allows the user (in this case, the user refers to administrators who are permitted to view the data) to search for information based on survey name to examine how various respondents answered questions for a certain survey. The search option also allows the user to search for a specific respondent's name, surname, or ID number to see how they responded to the various questions, allowing the well-being of the child to be assessed and appropriate steps to be taken.

## Data collection and analysis

Driven by their passion for improving lives, the professors carefully designed a two-phase study that would unfold over several years. In phase 1, they set out to collect baseline data using the CWTT app from a diverse group of children from five schools around Johannesburg. CoP digitally recorded information about the children's families, socioeconomic backgrounds and early life experiences. They delved into the children's academic performance, emotional development and physical health, leaving no stone unturned in their quest for knowledge. Before phase 2, children with challenges are identified by the CoP and relevant intervention is provided.

As the sun rose on a new day, CoP social workers swung into action, just a few days after the completion of phase 1 data collection. They sat together, heads bowed in collective determination, planning interventions tailored to each child's unique needs. The Sihleng'imizi family strengthening program stepped forward, offering a warm embrace to families facing tough times. With open hearts, they extended counseling, education and resources to empower parents to nurture their children with love and understanding. Meanwhile, the South African Depression and Anxiety Group (SADAG) wrapped it's comforting arms around children struggling with emotional hardships. With their expertise in mental health, they provided counseling sessions and safe spaces for these young souls to express their feelings and fears.

For the children grappling with health challenges, the UJ Optometry Clinic and the Wits University Audiology Clinic became beacons of care. Armed with state-of-the-art equipment and compassion in their hearts, these clinics conducted thorough assessments and provided necessary medical interventions. Education was not forgotten either. Social workers collaborated with schools and educators to implement tailored learning support for those facing difficulties in their studies. They celebrated every small achievement, nurturing a sense of accomplishment and confidence in the children's hearts. The issue of nutrition was also addressed with vigor. Community gardens sprouted, providing fresh and nutritious produce for families in need. The social workers organised workshops on balanced diets and culinary skills, turning mealtimes into joyful celebrations of health and unity. After the well-planned and well-received interventions, CoP began phase 2 of data collection. The questions that were used in phase 1 were adopted for this phase. Data analysis began after the completion of phase 2.

## Meta-analysis

The engineering team, using the Statistical Package for the Social Sciences (SPSS), prepared the data for a meta-analysis of pre- and post-interventions. The purpose of this analysis was to evaluate the effectiveness of the intervention in dealing with different challenges that were faced by children and their families. In preparation for meta-analysis, the descriptive data analysis for each question or variable was performed to produce the mean, standard deviation and study size for both pre- and post-interventions. The descriptive data analysis was then used to initiate the meta-analysis. From the meta-analysis, a forest plot was produced. Similar to factor analysis, the meta-analysis was performed in clusters or domains. However, for this discussion, only the forest plot that illustrated the intervention discussed in the preceding section was provided.

## Conclusion

Several technical conversations were undertaken to establish high-level functional requirements for the CWTT software. Thereafter, the architecture of the CWTT, which included Microsoft Excel, PowerApps and SQL database, was developed. Using different design platforms, the prototype of the CWTT was designed with various pages. Upon consensus on the CWTT design, the development stage began and the Microsoft Agile development approach was adopted. Thus, regular check-ins and meetings during the development phase were conducted as per the subsection on meeting cadence. The overall software development life cycle took 70 days. A unit test regime was adopted to test the developed CWTT. The questionnaire setup was performed using the simple interface found in Microsoft Excel, and the survey was completed by the users on a mobile device that had a PowerApps container.

The study was conducted in two key parts: phase 1 and phase 2. The data from both phases was stored in the Microsoft SQL database on multiple tables. The data was normalised by relational database standards; hence, it could not be interpreted and analysed. In preparation for phase 1 data analysis, the engineering team linked several tables of the database using SQL query commands to produce a single table. The data from the new table could be interpreted. The engineering team used SQL query commands to interpret the data and display the information on the graphic user interface. From the graphic user interface, the social workers identified and tracked children with different challenges. Thereafter, the social workers, together with other stakeholders, intervened to help children and families.

Factor analysis for both phase 1 and phase 2 was done using SPSS. The coding of the entire data set was automated and conducted in Excel. The 162 by 186 matrix data was imported into SPSS. As expected, the data converged on SPSS. The engineering team then uploaded each of the instrument's data into SPSS, hoping to secure convergence for explorative factor analysis. The data converged, but the statistical confidence of the results was very low. A similar procedure was adopted when dealing with the data set of the second phase. A meta-analysis was conducted to evaluate the effectiveness of the interventions conducted by social workers. The analysis found that most of the interventions were effective, especially in the areas of nutrition, education and home.

The CWTT is a user-friendly platform. The platform is open source; thus, in the future, it can be used and adopted by researchers interested in childhood development. It has offline capabilities, which allow the platform to be suitable for surveys conducted in remote areas. Geofencing allows the researchers to survey the individuals who reside in informal settlements.

## Exhibits

Figure 6.1 The CWTT architecture. Courtesy of University of Johannesburg

Figure 6.2 Landing page. Courtesy of University of Johannesburg

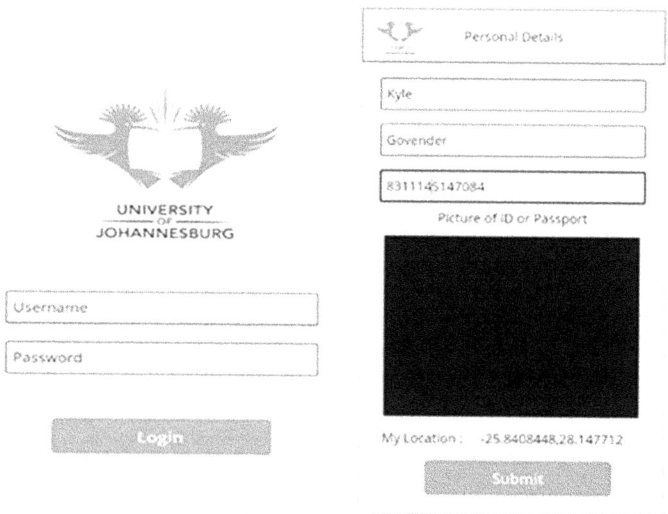

Figure 6.3 The UI of single selection questions. Courtesy of University of Johannesburg

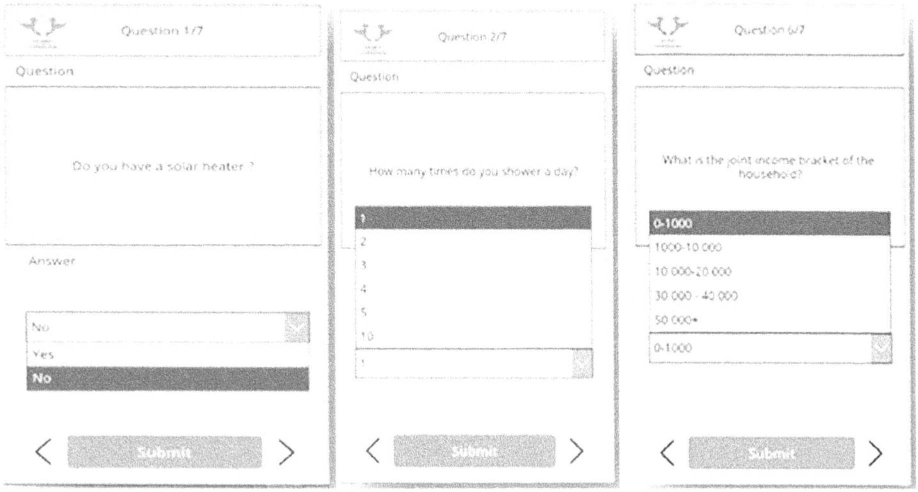


Figure 6.4 The UI of the date picker, free text question and multiple selection question. Courtesy of University of Johannesburg

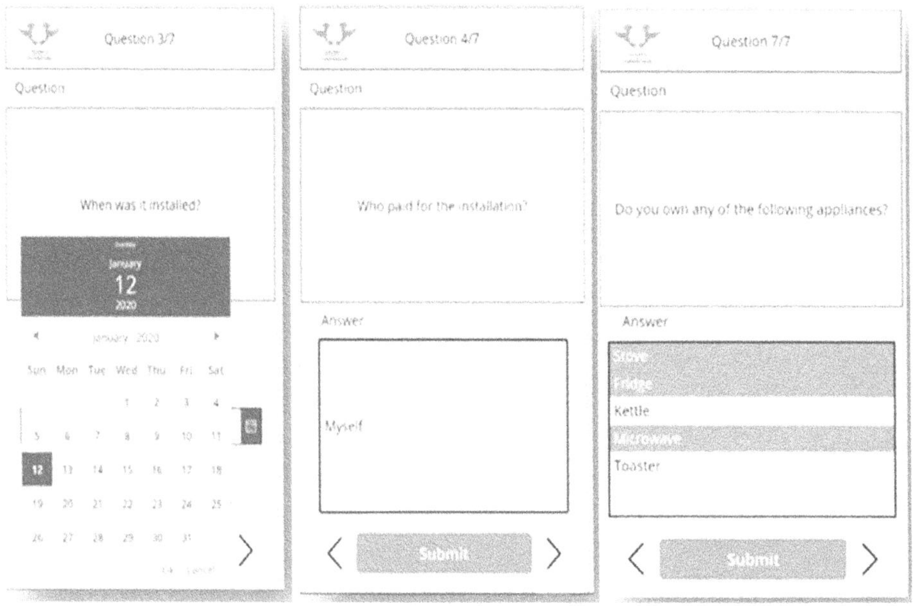

Figure 6.5 A screenshot of Microsoft Excel the user interface. Courtesy of University of Johannesburg

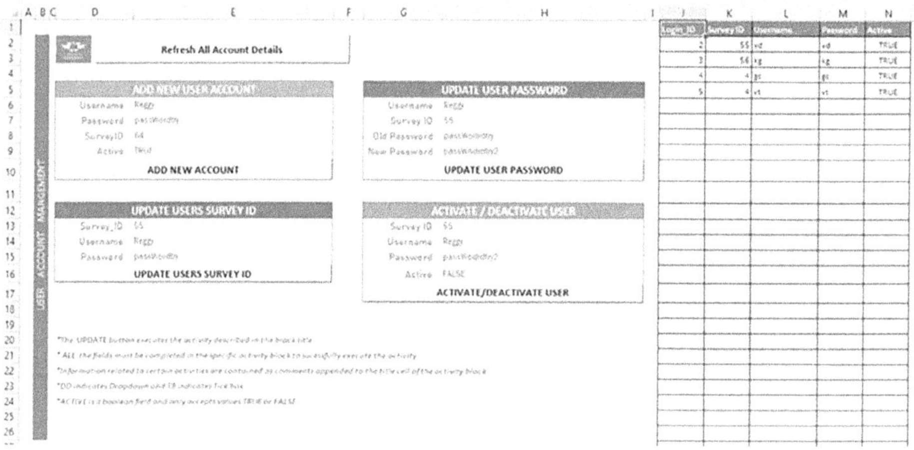


Figure 6.6 Merged results from different SQL tables (author's own)

| Responses_ID | Response | Survey_ID | Question_ID | NewGUID |
|---|---|---|---|---|
| 91 | Amazing,Good. | 1 | 3 | A1853890-2F25-4FD8-965C-51C8C2765138 |
| 92 | No | 1 | 4 | A1853890-2F25-4FD8-965C-51C8C2765138 |
| 93 | Skipped | 1 | 1 | 58B494C0-08B7-4F1D-9663-42196C58A667 |
| 94 | Skipped | 1 | 2 | 58B494C0-08B7-4F1D-9663-42196C58A667 |
| 95 | Something Different - Good. | 1 | 3 | 58B494C0-08B7-4F1D-9663-42196C58A667 |
| 96 | Skipped | 1 | 4 | 58B494C0-08B7-4F1D-9663-42196C58A667 |
| 97 | | 2 | 5 | 2F5783B0-D6F1-466A-9498-3A1778A66570 |
| 98 | 22 June 2021 | 2 | 6 | 2F5783B0-D6F1-466A-9498-3A1778A66570 |

Figure 6.7 A snapshot of the GUI, showing the data as presented in Figure 6 from the microsoft SQL database (author's own)

100 %

Results Messages

| | SurveyName | Question_Text | Response | Name | Surname | ID_Number | GPS_Location |
|---|---|---|---|---|---|---|---|
| 256 | Youth Behaviour | What is their age? | 12 | Testname1 | Testsu | Passporttest1 | -26.11834513 , 27.97086062 |
| 257 | Youth Behaviour | Are they currently studying? | Yes | Testname1 | Testsu | Passporttest1 | -26.11834513 , 27.97086062 |
| 258 | Youth Behaviour | What is their latest qualifications? | Degree Certificate | Testname1 | Testsu | Passporttest1 | -26.11834513 , 27.97086062 |
| 259 | Youth Behaviour | What is their Employment Status? | Unemployed Self-emplo... | Testname1 | Testsu | Passporttest1 | -26.11834513 , 27.97086062 |
| 260 | Youth Behaviour | Please provide the name of the person being interviewed. | Interview Two | Testname2 | Testsu | 1234567890123 | -26.11832614 , 27.97084349 |
| 261 | Youth Behaviour | What is their age? | 22 | Testname2 | Testsu | 1234567890123 | -26.11832614 , 27.97084349 |
| 262 | Youth Behaviour | Are they currently studying? | Yes | Testname2 | Testsu | 1234567890123 | -26.11832614 , 27.97084349 |
| 263 | Youth Behaviour | What is their latest qualifications? | Matric Certificate | Testname2 | Testsu | 1234567890123 | -26.11832614 , 27.97084349 |

Query executed successfully. DESKTOP-2356QSV\SQLEXPRESS ... DESKTOP-2356QSV\Xolani...

## Teaching notes

### Synopsis

A group of five SARChI chairs faced a challenge when they realised they needed to track the well-being of children in developing countries. They plan to use six key indicators to do so. However, they knew that collecting data for 163 children through a paper-based method would be difficult and could result in poor-quality data, redundancy and wasted time. The chairs also knew that their multidisciplinary studies would require more manpower from various disciplines, which would bring new challenges. To address these issues, the chairs formed a CoP to bring together all the necessary skills, including those from the engineering field. Although they were concerned about the financial implications of adopting a digital data collection method, they acknowledged that it may be the best option. By doing so, they hope to better understand and mitigate the challenges faced by children in developing countries.

### Teaching objectives

The case is written for use at the Johannesburg Business School (JBS) MBA level. The focus of the case is to understand and explore the challenges faced by social scientists when collecting data on children's well-being. The case will have guidelines for social scientists and CoPs on how they can track children's development across different domains:

- to analyse the importance of data collection in social studies
- to understand and explore the challenges faced by social scientists when collecting data
- to understand and explore the children's well-being domains
- to appraise the leadership and management skills portrayed in this case
- to analyse the importance of understanding your environment and context when making business decisions.

### Pedagogy

- Various concepts are addressed in this case, of which the most important are: data collection and analysis; exploring and mitigating challenges associated with data collection; leadership and management; diversification; digitalisation; and cloud computing.
- Solving the problems related to the case is not related to a right or wrong answer. Students should take each concept addressed in the case and identify the solutions that the main character applied in the case.
- Students can generalise the solutions from this case to any business type and any size of the organisation, leading to the sustainability of a sound business.

### Research methods

The case arises from an observation. The authors observed how CoP was formed with the intent of bringing in the necessary skills needed to build the foundation for the mitigation of the challenges faced by children. Further, the authors were involved in the development of the CWTT. The development adopted the following steps:

- user requirements specification
- prototyping

- agile development
- testing.

## Assignment questions

- What are the domains for children's well-being?
- What are the challenges faced by social scientists while collecting data?
- How do you track children's well-being using a digital platform?
- What is the effective way to collect data across all domains?

## The subject code and keywords

**Keywords:** Digital Data Collection; South African Research Chairs Initiative (SARChI); Community of Practice (CoP); Well-being Domains; Social Scientist; Interventions.

## Teaching plan and timing

- *10 min:* Class introduction: The case can be distributed to the class as prereading. One of the SARChI chairs should come to class and do the introduction from the heart.
- *30 min*: Assignment question 1: The instructor should get the students to work in pairs discussing the domains for children's well-being.
- *15 min*: Assignment question 2: The facilitator should request that the students identify three challenges faced by social scientists while collecting data.
- *15 min:* Assignment question 3: The facilitator should request that one group present the findings or recommendations related to tracking children's well-being using a digital platform.
- *10 min*: Assignment question 4: The facilitator should request that the students identify the effective way to collect data across all the domains.
- *5 min*: Conclusion: The instructor should wrap up the lesson by reflecting on the major challenges that the students identified and their recommendations.
- As it is a story being told, this case should be handled through discussion rather than PowerPoint presentations.

## Additional materials

The following links are recommended for more information.

http://ujtsuvey-912306333.af-south-1.elb.amazonaws.com:8000/
https://www.uj.ac.za/news/communities-of-practice-for-child-wellbeingproject-nears-completion/

## Analysis of assignment questions

A sample response of roughly 250 words should be provided for each assignment question. The application of model's frameworks and concepts from textbooks and journal articles should be included in the response to deepen the analysis and demonstrate understanding of current academic thinking in the area of study. The recommended core readings listed above should be used along with other resources.

## Key learnings

The instructor should wrap up the lesson by reflecting on the major challenges that the students identified or could not identify. Further, the instructors should provide the lessons with the solutions that they believe are appropriate for assignment questions as a form of conclusion.

### REFERENCES

Goldfeld S and Yousafzai A (2018) Monitoring tools for child development: an opportunity for action. *The Lancet Global Health* **6**(**3**): 232–233.

Snyder WM, Wenger E and de Sousa Briggs X (2004) *Community of Practice* in government: Leveraging knowledge for performance. *Public Manager* **32**(**4**): 17–22.

Vincent K, Steynor A, Waagsaether K and Culla T (2018) *Community of Practice*: One size does not fit all. *Climate Services* **11**: 72–77.

Wenger E, McDermott R and Snyder W (2002) *Cultivating Community of Practice*. Harvard Business School Press, Boston, MA, USA.

Arnesh Telukdarie
ISBN 978-1-83608-365-8
https://doi.org/10.1108/978-1-83608-362-720241009


# Chapter 7
# Defining digital transformation road map for an industrial organisation

**A Telukdarie and M Sishi**

## Introduction

It was in the latter part of the year 2022 when a new chief information officer (CIO) was appointed to take over the reins as the head of information and communications technology (ICT) for an industrial organisation that had been operating for over a century. Although the organisation had been profitable for the previous five consecutive years, most of its operational processes were manual, with only a few small digital systems working in silos without proper integration and collaboration. This gap would impact the organisation's profitability in the long run, as the absence of digital transformation deprives organisations of the real-time flow of information, efficiencies, visibility, transparency, performance management, informed decision-making and many more benefits. Therefore, the new CIO's assignment was to initiate the digital transformation of the organisation into an intuitive enterprise. This was going to be achieved by successfully developing a digital transformation roadmap that was aimed at transforming the organisation from a traditional (mixture of second and third Industrial Revolutions) to a digital and intuitive business (fourth Industrial Revolution). The CIO had to determine various aspects that would help in the development of a successful digital transformation roadmap: significant contributors to digital transformation, digital transformation approach and attributes of digitally intuitive business as key success factors.

## The background of Company X

Company X had many complex plant operations and made use of various manufacturing and expert systems in the handling of business-critical data across all business functions as well as for generating information and reports for managing performance and decision-making. These systems included business applications for functions such as accounting, materials management, plant maintenance and safety management, among others. Overall, the organisation was in the relatively early stages of digital technology adoption. Some business functions were using digital systems to some extent, while other business functions were still operating mostly manually.

Plant reliability, performance management and the flexibility to adapt to fluctuating input volumes while maintaining efficiency were critical factors in ensuring that Company X continued to deliver on its mission to create value in the African economy through the beneficiation of its by-product products. Competent management of these factors was critical in ensuring that the business continued to give the best possible value to its shareholders and stakeholders, including its strategic business partners.

It is typically the case that independent business operations implement stand-alone systems at the onset of digital adoption to serve their specific needs. However, as additional business functions adopt digital systems, these independent systems tend to perform complementary and sometimes duplicate functions, thereby making it vital for businesses to integrate across these standalone systems. However, the resultant 'spaghetti' integration is inefficient, has high in-service management costs, is unscalable and is prone to failure.

The need to configure automated data capture from operational technology (OT) hardware, namely operating equipment, analytical instruments and other devices such as RFID scanners, adds another layer of complexity to the integration scenario above. Although it is possible to create a data warehouse, wherein all data from the various functions may be pulled into and used to build customised reports, the reporting is not real-time, system performance is often slow and institutional memory is a challenge. Other challenges included manual operational interdependency, operations operating in isolation, manual inventory management, nonalignment of planning and scheduling, manual master data management and material variances picked up in the enterprise resource planning (ERP) system. Overall, digital transformation through standalone digital solutions tends to produce a business experience that is disjointed rather than seamless.

While targeted digital innovation could be used to build a more responsive enterprise, with integration among the various business functions, there was a need to transition the business from the reactionary enterprise phase. The operations included mostly isolated pockets of digital enablement, intending to migrate to an integrated and subsequently truly intuitive enterprise with embedded enterprise-wide analytics. An intuitive enterprise uses a user-centric design paradigm to build intelligent and flexible systems that are not only able to respond to changes in the market environment but can even pre-emptively sense and shape those markets.

The CIO employed a team of engineering researchers from the University of Johannesburg led by a seasoned professor, Dr. Arnesh Telukdarie, to fulfil the digital transformation roadmap mandate. The team embarked on a knowledge-based journey to gather information that would help in acquiring deeper knowledge pertinent to digital transformation, associated dependencies and supporting structures. The information gathered is captured in the following sections.

## Significant contributors to digital transformation

### Business strategy

The CIO and the research team observed business strategy definition as a set of inclusive plans used by an organisation to realise specific objectives (Olson *et al.*, 2021). Meskendahl (2010) defines business strategy as a mechanism to describe a firm's methods to compete with the players in the market. Meskendahl further states that the strategic approach describes the organisation's position towards performance and behaviour. The application of each business strategy is complex when combined with corporate strategy. To be considered successful, strategies should result in business growth, competitive advantage and strong financial performance (Sishi and Telukdarie, 2021). In 2008, Vodafone used a business strategy to close the gap with its leading competitor (TIM) by expanding its customer base and increasing cross- and up-selling to existing customers (Ghezzi, 2013). A business strategy encompasses one or more objectives, as illustrated in Table 7.1 and Table 7.2 (Heshmatisafa and Seppänen, 2023). The definition of each business strategy and related objectives is based on the organisation's current operational models.

## Digital business strategy

The CIO and the research team had to unearth similarities, differences and dependencies between traditional business strategy and digital business strategy. Therefore, various definitions of digital business strategy were explored. Businesses have been gradually transforming through digital technologies to achieve increased integration of systems, products, processes and services (Rafael *et al.*, 2020). Digital technologies were defined as a set of information, computing, communication and connectivity technologies. In their research towards the next generation of digital business strategy insights, Bharadwaj *et al.* (2013) identified four key themes: the scope of digital business strategy, the scale of digital business strategy, the speed of digital business strategy, and the sources of business value creation and capture in digital business strategy. These themes were aimed at capturing the characteristics of digital business strategy and expressing its nuances. A research study by da Costa *et al.* (2022) claimed that traditional business strategies have been radically reshaped by digital technologies as modular, dispersed, cross-functional and global business processes that enable work to be carried out across boundaries of time, distance and function. A digital business strategy is different from a traditional business strategy that is based on the current operational paradigms (Tutak, 2022). A digital business is transfunctional as it surpasses traditional functional areas and various business processes that are supported by ICT (Perera *et al.*, 2023; Rafael *et al.*, 2020). Moreover, a digital business strategy considers all significant contributors, such as technical architecture, change management, strategic roll-out plans and skills migration plans, together with advanced digital tools (Grooss *et al.*, 2022).

## Digital transformation

Digital transformation (DT) was observed as the application of new digital technologies for the enablement of significant business improvements (Zhai *et al.*, 2022). DT has become essential in strategic Information Systems (IS) research as well as for practitioners, as it uses digital technologies to drive significant changes in society and industries (Vial, 2019). Vial further claims that various organisations have been striving to find innovative options to use digital technologies by inventing strategies for embracing digital transformation and driving enhanced operational performance. It was discovered that all sectors of society, particularly economies, are significantly affected by DT. This presents sectors with opportunities to use such digital technologies as mobility, social networks, big data, the internet of things (IoT) and many more to radically change or improve their operating models (Singh *et al.*, 2020; Ziyadin *et al.*, 2020). This requires changes in core business operations, products, processes and organisational structures (Singh *et al.*, 2020). Reis *et al.* (2018) call on business leaders to adapt business strategies to an emerging digital paradigm. A DT framework by Vial (2019) employs digital technologies to yield disruptions that trigger strategic responses from organisations that endeavour to adjust their value creation course while managing structural changes and organisational impediments that influence positive and negative outcomes.

## The integration layers of the fourth industrial revolution

It was pivotal for the CIO and the research team to explore the fourth industrial revolution (4IR) architecture to transition the organisation from a traditional to a digital and intuitive business. Therefore, extensive knowledge of the 4IR architecture was researched and captured accordingly. In summary, 4IR encompasses three integration layers: vertical integration, end-to-end integration and horizontal integration (Sishi and Telukdarie, 2020; Zezulka *et al.*, 2016). The vertical integration connects the shop floor equipment and control systems with operational systems and enterprise-level systems and functions. End-to-end integration enables product lifecycle management, where a product's genealogy is developed by tracking a product from its inception through various stages in the vertical integration process, from the customer up to the end of life. The horizontal integration enables the organisation to collaborate with business partners such as suppliers, branches and customers, thus creating an integrated value chain network. The integration layers are illustrated in Figure 7.1.

## Developing the digital business strategy

Gathering information pertinent to the significant contributors to digital transformation equipped the CIO and research team with valuable information to develop a solid technique and process for developing a digital business strategy.

### Digital transformation approach

The key business mandate of the CIO was a digital business strategy outline with functional analysis and a forecast of a technological roadmap towards the delivery of an intuitive enterprise. Another key output was the implementation strategy. Accordingly, a structured approach was used to ensure the delivery of these output objectives, as illustrated in Figure 7.2.

The 1st phase of the digital transformation process entailed carrying out onsite and virtual information-gathering workshops and interviews. These workshops and interviews were used to clarify the existing business operations, functions, processes and resources. The second phase entailed the review and analysis of the information gathered during the 1st phase. This phase required a technical evaluation of installed systems and as-is operations, functions, processes and resources. The third phase entailed the use of information analysed in the second phase to document the as-is and to-be status from digital maturity and transformation perspectives. The fourth phase finalised and documented the digital business strategy. The technical sessions were conducted in this phase and the research team presented their findings and documented processes to the relevant process owners and business leaders to verify and validate.

Follow-up meetings with system experts were scheduled as required to finalise any outstanding technical details. After careful comparisons of business goals and user needs and specifications against ICT and operational technology (OT) infrastructure, the market was researched for cutting-edge hardware and software solutions available as options, and recommendations were made regarding possible ICT and OT infrastructure upgrades as well as other hardware requirements. The research team consolidated the gathered information and commenced with the documentation of business requirements and technical information. The as-is and to-be sign-off was achieved by circulating, presenting and discussing the digital business strategy document with the relevant business and technical stakeholders from each of the business processes and systems being interfaced.

## Information gathering and analysis

The onsite and virtual information gathering workshops and interviews were conducted with stakeholders responsible for the various areas of the business: operations, maintenance management, master data, finance, sourcing and logistics, quality, laboratory, safety, health environment, security, and commercial and operational technology. These sessions enabled the technical evaluation of currently installed systems, business operations, functions, processes and resources.

## Documentation: as-is and to-be statuses

Information gathering and analysis phases enabled the documentation of the respective as-is and the required future (to-be) statuses of Company X, which are detailed in an earlier section that provides the company background (digital transformation gaps and opportunities).

## Digital business strategy: finalisation and documentation

An important consideration in the design of a digital business strategy is the rapid evolution of technology. Global competitiveness can only be retained if optimisation is achieved and sustained. The digital business strategy proposed by the research team included total business visibility, delivering on multidimensional visibility and workflows enabling optimisation. The strategy would be sustained by the suggested development of internal technical capacities with long-term research and development support, including artificial intelligence (AI). The key approach was achieving the current best practice apex (integrated digital business) and developing a secondary apex envisaging the future (moving towards an intuitive business). The strategy included the following components.

- Intuitive layer (upper apex): This layer is dependent on an integrated foundation of systems, supported by business intelligence (BI) tools and intended to deliver value based on AI and analytics.
- Integrated ERP: Here, the researchers sought to achieve a collection of enterprise systems addressing the medium- to long-term operations of the business. This included business reporting and automated workflows, driving total business visibility and facilitating automation.
- Integration layer: This layer is an intermediate layer between ERP and all plant operations. Total integration of all data from enterprise and operations.
- Manufacturing layer: The entire manufacturing operation is executed by way of systems. Integration into the ERP by way of the data layer maximises operations. All plant data is integrated into this layer.
- Plant control network (PCN): Plant sensors, controllers, switchers, advanced control and data storage are networked to provide bottom-up data.

The ability to implement technical systems, capacitate the team, develop service management and still operate a complex facility brings significant risk. Therefore, this digital business strategy dictated business case development with a prioritisation strategy that would focus on structured enablement while minimising risk. Developing technical solutions commences with the definition of roadmaps, transitioning into a plan, following a design and a market evaluation process. Immediate, impulsive and total digitalisation was not going to be a viable option. Therefore, the research team recommended a stage-wise approach, commencing with the integration or data layer and identifying strategic and priority systems with purposeful execution. The development of key digital support structures such as a central management database (CMDB), service level agreements (SLAs), comprehensive architecture capacities and preparing the company by developing digital skills were essential.

A first go-to recommendation was that change management be strategically defined as a priority. The ability to align the business strategy with the digital strategy was essential. Executive heads should support the digital strategy and its enablement. Executive time would secure buy-in from team members. Executive awareness would raise team awareness and facilitate delivery. Technical solutions should be supported at all levels, as they would fail if the company's strategies did not align with digital strategies. This, together with appropriately equipped technical and nontechnical teams, should be planned for. Technology is highly people-centric; thus, the teams must be aware of and comfortable with the change. The key was to also begin/start with reskilling strategies as part of change management. Herein, a skills migration strategy was proposed as a lead consideration. The research team suggested that the digital business strategy components identified earlier in this section should be arranged hierarchically with the following layers.

- Strategy layer: Digital enablement and visualisation of data and advanced AI tools for advancing an intuitive business.
- ERP layer: The installation of various additional modules, the upgrading of the current modules and a fully integrated ERP landscape.
- Data layer: A fully integrated ERP and operational data layer was proposed. All company data should be available and form the basis for BI. BI should be configured for scheduled reporting and ad hoc requests.
- Manufacturing Enterprise Solution (MES) layer: A fully functional MES layer would facilitate ERP enablement and operational execution efficiency. MES data collection by way of a historian was deemed important. Operational mobility was deemed the low-cost solution for operational data in older facilities.
- PCN layer: The control layer was identified as the data source for an integrated business. Various systems to ensure consistent and reliable data were proposed.

Executive strategy and quantification should be based on some fundamentals. The key was to drive an intuitive enterprise, with the first step being the delivery of a responsive enterprise. As depicted in Figure 7.3, this digital business strategy was built on five key pillars: *(i)* strategy, *(ii)* digital culture, *(iii)* digital hardware, *(iv)* data and systems maturity, and *(v)* digital service management to facilitate digital maturity. These pillars, if enabled, would strategically deliver significant value.

Having identified that the organisation had limited digital operations and notable opportunities in data and systems integrations, the researchers and the CIO jointly finalised the digital business strategy that outlined how the company may put in place systems to ensure full digital integration across all business functions. Based on the landscaping and global best practices, a high-level architecture was proposed, as depicted in Figure 7.4. This high-level architecture encompasses five levels: *(i)* the ERP level with various enterprise functions, *(ii)* the business integration level that connects operational systems with the ERP, *(iii)* the operational level encompassing various operational systems, *(iv)* the integration level encompassing three key aspects: historian, data warehouse and PCN responsible for linking all operational equipment, and *(v)* the plant automation level encompassing operational equipment and control systems. The assimilation of these five levels would create a connected enterprise that comprises both the vertical and end-to-end integration components of 4IR. This architecture further enabled the third component of 4IR, horizontal integration, to create an integrated value chain network by incorporating suppliers, various branches and customers. The BI component would extract and report on company-wide integrated data and information, while the AI component would usher the organisation into an intuitive business using structured and unstructured data.

This high-level architecture was based on the details in the fully digital business strategy document. An analysis of as-is business processes and systems was detailed with an outline of a strategy for the implementation of integrated digital business systems, with details on technological requirements, functional elements, constraints and opportunities. The outline contained actionable details of how the business may transform into a fully self-optimising, intuitive operation and highlighted all milestones that should be attained during this transformation. The document specified systems functionalities, integration, data exchange specifications and architecture, as well as functional and cross-functional key performance indicators (KPIs). Furthermore, the digital business strategy document served as a statement of requirements following the extraction of relevant data through multiple site visits, technical sessions and consultant-client workshops.

## Key benefits (attributes of digitally intuitive business)

The full enablement of this strategy was based on the current operations and operational challenges, together with the current digital tools and systems available. A high-level business benefits analysis of this digital business strategy included, but was not limited to:

- Total business visibility.
- Automation of repetitive, non-value-adding activities. This included event-based automation, criterion-based automation and interdependent automation.
- Enhanced data collection.
- Energy, utility and raw material digitalisation.
- CMDB: The ability to document all assets, hardware and software was critical. The mastery of change, design and safety was essential to eradicating errors, optimising operations and facilitating service management and transition. CMDB was deemed a safety net for change.
- Process mapping which includes business activities, time taken and resources utilised should be mapped, which would serve as the organisational memory. The maps were also excellent inputs for optimisation and the design of systems.
- Machine-based inspections that use machine learning (ML) and AI present the opportunity to overcome subjective interpretations of visual standards for defect determination. The business benefits were improved reliability and higher throughput.
- Reduction of manual capture or recording of data would lead to reduced human error and an increase in production efficiency. This would result in fewer mistakes reaching the market and therefore fewer customer complaints.

- Digital product development tools incorporating digital twins could allow for the simulation of the product development process in a virtual environment with convenient testing of different process parameter values, thereby leading to early identification and resolution of challenges and overall quicker product development times.
- Nonconformances could be logged into a system and automatically routed to relevant actioners.

## Lessons for the CIO and research team

The lessons learned by the CIO and research team suggested that most companies, especially in developing countries, are not digitised, while some are trying to adopt digital transformation through the digitalisation and integration of stand-alone systems. This results in complex and unstructured integrations that are inefficient, difficult and costly to manage, unscalable and susceptible to failure. The impetuous adoption of digital transformation is often not aligned with the overall business strategy. The output of this work introduced an effective approach to defining a roadmap for digital transformation. The output revealed that the business strategy is based on operational paradigms, while the digital business strategy considers all significant contributors, including technical architecture, change management, a strategic roll-out plan and a skills migration plan, together with advanced digital tools. Furthermore, the output emphasised the importance of aligning the digital business strategy with the overall business strategy.

The approach to developing the digital business strategy should be based on detailed landscaping of the business systems, operations, functions, processes and resources. This should be followed by discovering the industry's best practices and the company's existing business strategy. The digital business strategy should then be defined and aligned to the overall business strategy to transform (through optimisation), grow (through replication) and sustain (through product diversification and empowerment) the business. The digital business strategy should strive to achieve intuitive business by first creating a responsive enterprise. The digital business strategy should be maintained through the development of internal technical capacities, together with long-term research and development support and AI.

It was strongly suggested that change management be strategically defined as one of the top priorities. Change management should encompass, among other important things, the development of digital skills. It was further suggested that executive support of the digital business strategy and its enablement is pivotal. This will help the digital business strategy secure business buy-in and ensure that technical solutions are supported at all levels and are aligned with the overall business strategy.

The high-level architecture for the digital business strategy would digitally transform the organisation by implementing vertical, horizontal and end-to-end integration components of 4IR. This would yield a connected enterprise that realises a real-time flow of information throughout the organisation, product genealogy and connected world immersion. The incorporation of BI and AI would enable efficient reporting from integrated data and information and intuitive business using structured and unstructured data. The details of the approach to implementing the digital strategy were stipulated in the full digital business strategy document. The details included items such as technological requirements, functional elements, constraints, opportunities, systems functionalities, integration, data exchange specifications, architecture, functional and cross-functional KPIs, and many more. The next challenge for the CIO and Company X would be the practical implementation of the digital business strategy. The results of the implementation would reveal whether the digital business strategy is effective and would yield the suggested business benefits.

## Exhibits

Table 7.1 Business strategies with associated objectives

| Business Strategy | Objective |
|---|---|
| Financial | Maintenance of adequate working capital |
| Marketing | Growth and maintenance of leading market share |
| Operational | Improvement of efficiency through automation |

Figure 7.1 The fourth industrial revolution integration layers (author's own)

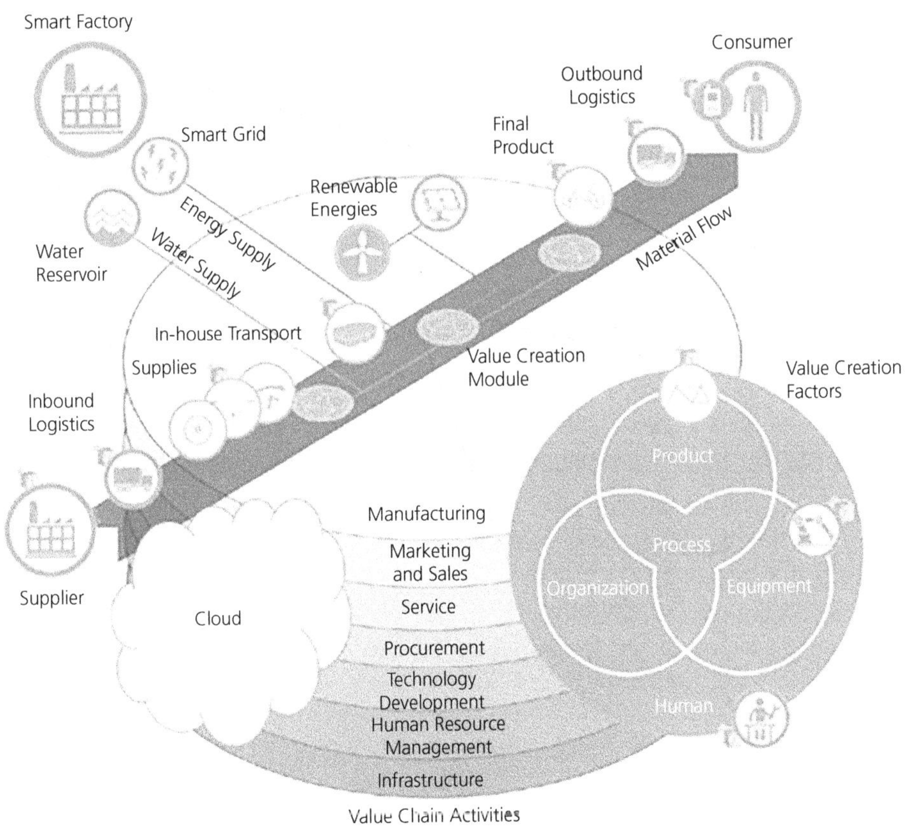

Figure 7.2 Followed to develop the digital business strategy (author's own)

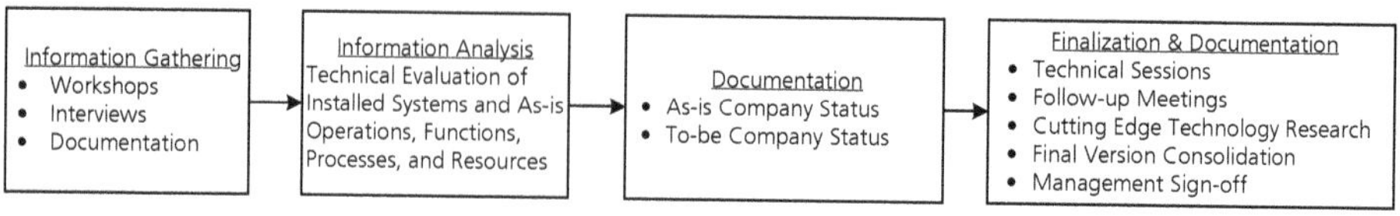


Table 7.2 Mapping of business functional areas to S/4 HANNA LOB modules (author's own)

| Business Function | S/4 HANA LOB Module |
|---|---|
| Operations | Manufacturing Management, Analytics, Asset Care and Production Planning. |
| Accounting and Finance | Finance |
| Maintenance, Health Safety and Environment | Asset Care, Manufacturing Management, Research and Development/ Engineering. |
| Quality | Quality Management |
| Security | Access Control Management System |
| Logistics | Customer Service, Materials Management, Supply Chain Management |
| Warehousing | Manufacturing Management, Sourcing and Procurement |
| Human Resources | Human Capital Management, Human Resources |
| Sourcing | Supply Chain Management, Materials Management, Supplier Relationship Management |
| Treasury | Sales, Finance, Analytics |

Figure 7.3 Key pillars of digital business strategy (author's own)

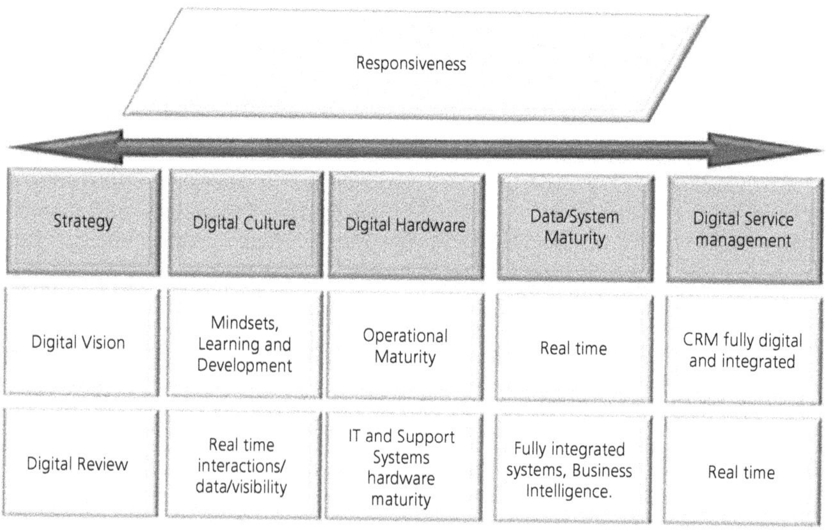

Figure 7.4 High-level digital business strategy architecture (author's own)

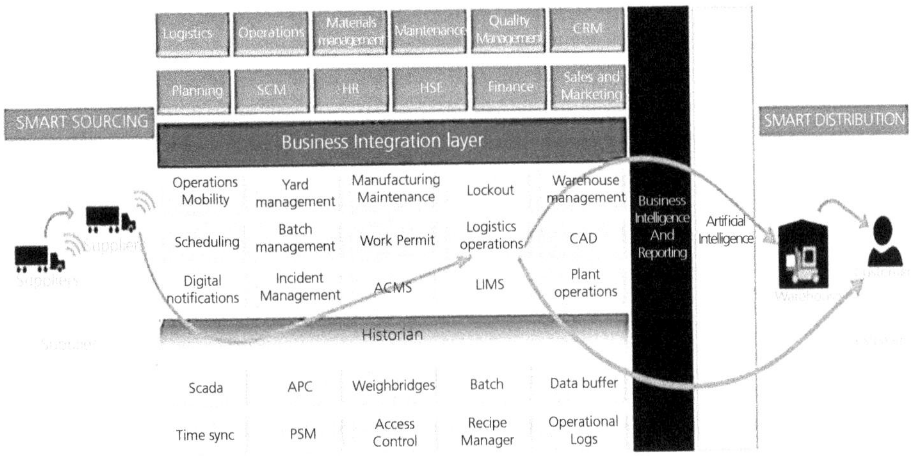


## Teaching notes

### Synopsis

It was in the latter part of the year 2022 when a new CIO was appointed to take over the reins as the head of information and communications technology (ICT) for an industrial organisation that had been operating for over a century. Even although the organisation had been profitable for the previous five consecutive years, most of its operational processes were manual, with only a few small digital systems working in silos without proper integration and collaboration. This gap would impact the organisation's profitability in the long run, as the absence of digital transformation deprives organisations of the real-time flow of information, efficiencies, visibility, transparency, performance management, informed decision-making and many more benefits. Therefore, the new CIO's assignment was to initiate the digital transformation of the organisation into an intuitive enterprise. This was going to be achieved by successfully developing a digital transformation roadmap that was aimed at transforming the organisation from a traditional (mixture of second industrial revolution and third industrial revolution) to a digitally intuitive business (4IR). The CIO had to determine various aspects that would help in the development of a successful digital transformation roadmap: significant contributors to digital transformation, digital transformation approach and attributes of digitally intuitive business as key success factors.

### Teaching objectives

This case study is intended for use in undergraduate, MBA and executive education courses in Strategy and Digital Transformation. The nature of the teaching objectives suggests that this case study is more appropriate for undergraduate programs or introductory strategy courses in postgraduate programs.

The case has five pedagogical objectives.

1. **Critical thinking**: To comprehend the strategic choices that companies need to make to achieve business objectives for growth and sustainability.
2. **Significant contributors**: To understand and leverage the significant contributors to the digital transformation of organisations, especially in developing countries.
3. **Digital transformation approach**: To understand possible approaches to digitally transforming industrial organisations.
4. **Key success factors**: To uncover attributes that are exhibited by digitally intuitive businesses to use them as a benchmark for the digitalisation of industrial organisations.

## Research methods

This case draws upon data, information and insights collected from onsite and virtual information-gathering interviews and workshops to understand the as-is and future state of the organisation. This was followed by extensive market research for cutting-edge hardware and software solutions available as options and recommendations were made regarding possible ICT and OT infrastructure upgrades and other hardware requirements. The as-is and to-be sign-off was achieved by documenting, circulating, presenting and discussing the digital business strategy document with the relevant business and technical stakeholders from each of the business processes and systems being interfaced.

## Keywords and subject code

Business Strategy, Digital Business Strategy, Digital Transformation and Digital Maturity

## Assignment questions

1. What are the strategic choices an organisation should make to achieve business objectives for business growth and sustainability?
2. What are the significant contributors to the digital transformation of organisations in developing countries?
3. What are effective approaches to transforming industrial organisations digitally?
4. What are the positive attributes that are exhibited by organisations that have transformed into digitally intuitive businesses? Should these attributes be used as drivers of the digital transformation of organisations?

## Further reading

The authors recommend the following reading.

Bharadwaj A, El Sawy OA, Pavlou PA and Venkatraman N (2013) Digital business strategy: toward a next generation of insights. *MIS Quarterly* **37(2)**: 471–482.

da Costa LS, Munhoz IP, Pereira L and Santos Akkari AC (2022) Assessing the digital maturity of micro and small enterprises: A focus on an emerging market. *Procedia Computer Science* **200**: 175–184.

Fields D (2022) Automated landlord: digital technologies and postcrisis financial accumulation. *Environment and Planning A: Economy and Space* **54(1)**: 160–181.

Kıyıklık A, Kuşakcı AO and Mbowe B (2022) A digital transformation maturity model for the airline industry with a self-assessment tool. *Decision Analytics Journal* **3**: 1–13.

Sishi M and Telukdarie A (2020). Implementation of Industry 4.0 technologies in the mining industry case study. *International Journal of Mining and Mineral Engineering* **11(1)**: 1–22.

## Teaching plan and timing

For a class of 90 minutes, the researcher recommended the following teaching plan.

- *5 min*: Introduction. The instructor provides a summary of the case.
- *20 min*: Question 1: What are the strategic choices an organisation should make to achieve business objectives for business growth and sustainability?
- *20 min*: Question 2: What are the significant contributors to the digital transformation of organisations in developing countries?
- *20 min*: Question 3: What are effective approaches to transforming industrial organisations digitally?
- *20 min*: Question 4. What are the positive attributes that are exhibited by organisations that have transformed into digitally intuitive businesses? Should these attributes be used as drivers of the digital transformation of organisations?
- *5 Minutes*: Conclusion. Wrap up the case and summarise key learnings.

## Analysis of assignment questions

**1. What are the strategic choices an organisation should make to achieve business objectives for business growth and sustainability?**

The value proposition of this question is to educate learners on the importance of business strategies and efficient techniques for devising these strategies.

Most resources define the business strategy as a high-level plan for a company to reach specific business objectives. Strategies are deemed successful when they yield business growth, a strong competitive advantage and strong financial performance. It is pivotal for a business strategy to succeed; otherwise, the company should change the strategy or prepare for going out of business.

A business strategy is also viewed as the organisation's working idea to realise its vision, prioritise objectives, compete productively and strengthen financial performance with its business model. The formulation of business strategy focuses on meeting objectives; therefore, most businesses begin/start strategising by focusing on the highest-level business objectives, such as increasing owner value (see Figure 7.5). This objective is realised by the firm's ability to earn profit. The business strategy subsequently aspires to earn, sustain and grow profit.

Business strategy is easily understood if each strategy is viewed as one element of a strategic framework. At the top of the strategic framework hierarchy sits top-level business strategy—that is, earn, sustain and grow profit. The lower level of the hierarchy encompasses strategies that describe how the organisation achieves the top-level strategy. Examples of lower-level strategies and objectives are illustrated in Table 7.3.

Table 7.3 Lower-level strategies and objectives

| Lower-level strategy | Objective |
|---|---|
| Financial | Maintenance of adequate working capital |
| Marketing | Grow and maintain a leading market share |
| Operational | Improve efficiency through automation |

*[The teaching suggestion]* The instructor can ask students to devise a generic business strategy, low-level business strategies and business objectives for each low-level strategy.

One of the possible techniques for devising a business strategy is to place the generic strategy at the top of the strategic framework. The generic strategy should be underpinned by multiple low-level strategies such as marketing strategy, operational strategy, financial strategy and so forth. The marketing strategy can encompass business objectives such as sales revenue, market share, branding, product differentiation, customer demand, pricing and many more. The business objectives for operational strategy would be operational efficiency, inventory turnover, just in time (JIT) manufacturing, improved production and fixed asset utilisation. The financial strategy can have, among others, business objectives such as capital structure leverage, financial structure leverage, cost of capital, working capital, pricing and return on assets. Table 7.1 depicts the potential solution.

**2. What are the significant contributors to the digital transformation of organisations in developing countries?**

*[The teaching suggestion]* The instructor can ask students to identify and provide a brief definition of each significant contributor to digital transformation that was identified by the research team hired by the CIO and led by a seasoned professor from the University of Johannesburg Digital Business School.

The digital transformation identified by the research team is as follows.

### Business strategy

The CIO and the research team observed business strategy definition as a set of inclusive plans used by an organisation to realise specific objectives. Many research studies define business strategy as a mechanism to describe a firm's methods to compete with the players in the market. The strategic approach describes the organisation's position towards performance and behaviour. The application of each business strategy is complex when combined with corporate strategy. To be considered successful, strategies should result in business growth, competitive advantage and strong financial performance. A reference can be made to Vodafone, which used a business strategy to close the gap with its leading competitor (TIM) by expanding its customer base and increasing cross- and up-selling to existing customers in 2008.

### Digital business strategy

The CIO and the research team had to unearth similarities, differences and dependencies between traditional business strategy and digital business strategy. Therefore, various definitions of digital business strategy were explored. Businesses have been gradually transforming through digital technologies to achieve increased integration of systems, products, processes and services. Digital technologies were defined as a set of information, computing, communication and connectivity technologies. In their research towards the next generation of digital business strategy insights, the authors identified four key themes: the scope of digital business strategy, the scale of digital business strategy, the speed of digital business strategy, and the sources of business value creation and capture in digital business strategy. These themes were aimed at capturing the characteristics of digital business strategy and expressing its nuances. Other research claims that traditional business strategies have been radically reshaped by digital technologies as modular, dispersed, cross-functional and global business processes that enable work to be carried out across boundaries of time, distance and function. A digital business strategy is different from a traditional business strategy that is based on the current operational paradigms. A digital business is trans functional as it surpasses traditional functional areas and various business processes that are supported by ICT. Moreover, a digital business strategy considers all significant contributors, such as technical architecture, change management, strategic roll-out plans and skills migration plans, together with advanced digital tools.

### Digital transformation

DT was observed as the application of new digital technologies for the enablement of significant business improvements. DT has become essential in strategic information systems (IS) research as well as for practitioners, as it uses digital technologies to drive significant changes in society and industries. Various organisations have been striving to find innovative options to use digital technologies by inventing strategies for embracing digital transformation and driving enhanced operational performance. It was discovered that all sectors of society, particularly economies, are significantly affected by DT. This presents sectors with opportunities to use such digital technologies as mobility, social networks, big data, the IoT and many more to radically change or improve their operating models. This requires changes in core business operations, products, processes and organisational structures. Business leaders are called to adapt business strategies to an emerging digital paradigm. DT frameworks employ digital technologies to yield disruptions that trigger strategic responses from organisations that endeavour to adjust their value creation course while managing structural changes and organisational impediments that influence positive and negative outcomes.

### The integration layers of the fourth industrial revolution

It was pivotal for the CIO and the research team to explore the 4IR architecture to transition the organisation from a traditional to a digital and intuitive business. Therefore, extensive knowledge of the 4IR architecture was researched and captured accordingly. In summary, 4IR encompasses three integration layers: vertical integration, end-to-end integration and horizontal integration. The vertical integration connects the shop floor equipment and control systems with operational systems and enterprise-level systems and functions. End-to-end integration enables product lifecycle management, where a product's genealogy is developed by tracking a product from its inception through various stages in the vertical integration process, from the customer, up to the end of life. The horizontal integration enables the organisation to collaborate with business partners such as suppliers, branches and customers, thus creating an integrated value chain network. The integration layers are illustrated in Figure 7.1.

### 3. What are effective approaches to transforming industrial organisations digitally?

The engineering team proposed a systematic approach to devise a digital business strategy and implementation methods for Company X. This approach and method were underpinned by extensive knowledge gathered on the significant contributors to the digital transformation of organisations.

*[The teaching suggestion]* The instructor can ask learners to identify and discuss an approach to devising a digital business strategy and implementation methods for companies that seek to become intuitive digital businesses.

The approach adopted by the engineering team to devising a digital business strategy for Company X is as follows.

#### Information gathering and analysis

The onsite and virtual information gathering workshops and interviews were conducted with stakeholders responsible for the various areas of the business: operations, maintenance management, master data, finance, sourcing and logistics, quality, laboratory, safety, health environment, security, commercial and operational technology. These sessions enabled the technical evaluation of currently installed systems, business operations, functions, processes and resources.

#### Documentation: as-is and to-be statuses

Information gathering and analysis phases enabled the documentation of the respective as-is and the required future (to-be) statuses of Company X.

#### Digital business strategy: finalisation and documentation

An important consideration in the design of a digital business strategy is the rapid evolution of technology. Global competitiveness can only be retained if optimisation is achieved and sustained. The digital business strategy proposed by the research team included total business visibility, delivering on multidimensional visibility and workflows enabling optimisation. The strategy would be sustained by the suggested development of internal technical capacities with long-term research and development support, including artificial intelligence (AI). The key approach was achieving the current best practice apex (integrated digital business) and developing a secondary apex envisaging the future (moving towards an intuitive business). The strategy included the following components.

- Intuitive layer (upper apex): This layer was dependent on an integrated foundation of systems, supported by BI tools and intended to deliver value based on AI and analytics.
- Integrated ERP: Here, the researchers sought to achieve a collection of enterprise systems addressing the medium- to long-term operations of the business. This included business reporting and automated workflows, driving total business visibility and facilitating automation.
- Integration layer: This layer is an intermediate layer between ERP and all plant operations. Total integration of all data from enterprise and operations.

- Manufacturing layer: The entire manufacturing operation is executed by way of systems. Integration into the ERP by way of the data layer maximises operations. All plant data is integrated into this layer.
- Plant control network (PCN): Plant sensors, controllers, switchers, advanced control and data storage are networked to provide bottom-up data.

The ability to implement technical systems, capacitate the team, develop service management and still operate a complex facility brings significant risk. Therefore, this digital business strategy dictated business case development with a prioritisation strategy that would focus on structured enablement while minimising risk. Developing technical solutions commences with the definition of roadmaps, transitioning into a plan, following a design and a market evaluation process. Immediate, impulsive and total digitalisation was not going to be a viable option. Therefore, the research team recommended a stage-wise approach, commencing with the integration or data layer and identifying strategic and priority systems with purposeful execution. The development of key digital support structures such as a central management database (CMDB), service level agreements (SLAs), comprehensive architecture capacities and preparing the company by developing digital skills were essential.

A first go-to recommendation was that change management be strategically defined as a priority. The ability to align the business strategy with the digital strategy was essential. Executive heads should support the digital strategy and its enablement. Executive time would secure buy-in from team members. Executive awareness would raise team awareness and facilitate delivery. Technical solutions should be supported at all levels, as they would fail if the company's strategies did not align with digital strategies. This, together with appropriately equipped technical and nontechnical teams, should be planned for. Technology is highly people-centric; thus, the teams must be aware of and comfortable with the change. The key was to also begin/start with reskilling strategies as part of change management. Herein, a skills migration strategy was proposed as a lead consideration. The research team suggested that the digital business strategy components identified earlier in this section should be arranged hierarchically with the following layers.

1. Strategy layer: Digital enablement and visualisation of data and advanced AI tools for advancing an intuitive business.
2. ERP layer: The installation of various additional modules, the upgrading of the current modules and a fully integrated ERP landscape.
3. Data layer: A fully integrated ERP and operational data layer was proposed. All company data should be available and form the basis for BI. BI should be configured for scheduled reporting and ad hoc requests.
4. Manufacturing enterprise solution (MES) layer: A fully functional MES layer would facilitate ERP enablement and operational execution efficiency. MES data collection by way of a historian was deemed important. Operational mobility was deemed the low-cost solution for operational data in older facilities.
5. PCN layer: The control layer was identified as the data source for an integrated business. Various systems to ensure consistent and reliable data were proposed.

Executive strategy and quantification should be based on some fundamentals. The key was to drive an intuitive enterprise, with the first step being the delivery of a responsive enterprise. As depicted in Figure 7.3, this digital business strategy was built on five key pillars: *(i)* strategy, *(ii)* digital culture, *(iii)* digital hardware, *(iv)* data and systems maturity, and *(v)* digital service management to facilitate digital maturity. These pillars, if enabled, would strategically deliver significant value.

Having identified that the organisation had limited digital operations and notable opportunities in data and systems integrations, the researchers and the CIO jointly finalised the digital business strategy that outlined how the company may put in place systems to ensure full digital integration across all business functions. Based on the landscaping and global best practices, a high-level architecture was proposed, as depicted in Figure 7.4. This high-level architecture encompasses five levels: *(i)* ERP level with various enterprise functions, *(ii)* business integration level that connects operational systems with the ERP, *(iii)* operational level encompassing various operational systems, *(iv)* integration level encompassing three key aspects: historian, data warehouse and PCN responsible for linking all operational equipment, and *(v)* the plant automation level encompassing operational equipment and control systems. The assimilation of these five levels would create a connected enterprise that comprises both the vertical and end-to-end integration components of 4IR. This architecture further enabled the third component of the 4IR, horizontal integration, to create an integrated value chain network by incorporating suppliers, various branches and customers. The BI component would extract and report on company-wide integrated data and information, while the AI component would usher the organisation into an intuitive business using structured and unstructured data.

This high-level architecture was based on the details in the fully digital business strategy document. An analysis of as-is business processes and systems was detailed with an outline of a strategy for the implementation of integrated digital business systems, with details on technological requirements, functional elements, constraints and opportunities. The outline contained actionable details of how the business may transform into a fully self-optimising, intuitive operation and highlighted all milestones that should be attained during this transformation. The document specified systems functionalities, integration, data exchange specifications and architecture, as well as functional and cross-functional key performance indicators (KPIs). Furthermore, the digital business strategy document served as a statement of requirements following the extraction of relevant data through multiple site visits, technical sessions and consultant-client workshops.

**4. What are the positive attributes that are exhibited by organisations that have transformed into digitally intuitive businesses? Should these attributes be used as drivers of the digital transformation of organisations?**

*[The teaching suggestion]* The instructor can ask learners to identify attributes exhibited by organisations that have been transformed into digitally intuitive businesses. Students can further suggest if these attributes can be utilised as a benchmark for the digital transformation of organisations.

A high-level business benefits analysis of this digital business strategy included, but was not limited to, the following.

- Total business visibility.
- Automation of repetitive, non-value-adding activities; this included event-based automation, criterion-based automation and interdependent automation.
- Enhanced data collection.
- Energy, utility and raw material digitalisation.
- CMDB: The ability to document all assets, hardware and software was critical. The mastery of change, design and safety was essential. This would eradicate errors, optimise operations and facilitate service management and transition. CMDB was deemed a safety net for change.
- Process mapping: The business activities, time taken and resources utilised should be mapped, which would serve as the organisational memory. The maps were also excellent inputs for optimisation and the design of systems.
- Machine-based inspections that use machine learning (ML) and AI presented the opportunity to overcome subjective interpretations of visual standards for defect determination. The business benefits were improved reliability and higher throughput.
- Reduction of manual capture or recording of data would lead to reduced human error and an increase in production efficiency. This would result in fewer mistakes reaching the market and therefore fewer customer complaints.
- Digital product development tools incorporating digital twins could allow for simulation of the product development process in a virtual environment with convenient testing of different process parameter values, thereby leading to early identification and resolution of challenges and overall quicker product development times.
- Nonconformances could be logged into a system and automatically routed to relevant actionees.
- *[The teaching suggestion]* Instructors could ask students to also debate whether the benefits are representative of most of the benefits of digital transformation. Therefore, the following attributes could also be added to the list to form a solid benchmark for the digital transformation of organisations.
  - Mass customisation. The fluctuating requirements of production orders are accommodated. Changes are allowed at any time. It is also possible to run low production volumes but still make a profit.
  - Agility. Different aspects such as time, quality and price are well managed by intelligent and self-configuring production systems.
  - Better methods. Factories can be optimised at different levels in real time on a case-by-case basis.
  - Improved data analysis. Large quantities of data provided by IoT devices can be used for analysis purposes. For example, machine vendors can use data collected from their machines to improve machines' efficiency and strengthen customer relationships. This can be a result of examining machines' behaviour at various times.
  - Proactive maintenance can be achieved by collecting performance data (e.g. engine hours, mileage and temperature) in real time and using this data to determine when equipment should be booked for service.

- Integrated supply chain. IoT will deliver information in real time, enabling manufacturers to better understand their supply chain. A manufacturing company can connect equipment to suppliers, enabling visibility on interdependencies, material flow and production cycle time.

## Key learnings

The instructor can conclude the case by highlighting the elements that are pivotal to the approach for devising a digital business strategy. Scrutinising existing business systems, operations, functions, processes and resources is the basis of digital business strategy. This should be followed by research on global best practices, technologies and the company's existing business strategy. The digital business strategy should then be constructed and aligned with the overall business strategy. The digital business strategy should transform, grow and sustain the business. The digital business strategy should begin with the delivery of a responsive enterprise and follow with an intuitive enterprise. Through change management, employees should be empowered with digital skills to adapt to new ways of doing things. Executives should support technical solutions and the overall digital strategy to enable buy-in from team members. The high-level digital business strategy architecture in Figure 7.4 ushers organisations into adopting the 4IR concepts and technologies. This would help an organisation transform into a connected enterprise and be part of the connected world.

## Exhibits

Figure 7.5 Strategic framework. Source: Sishi and Telukdarie (2021)

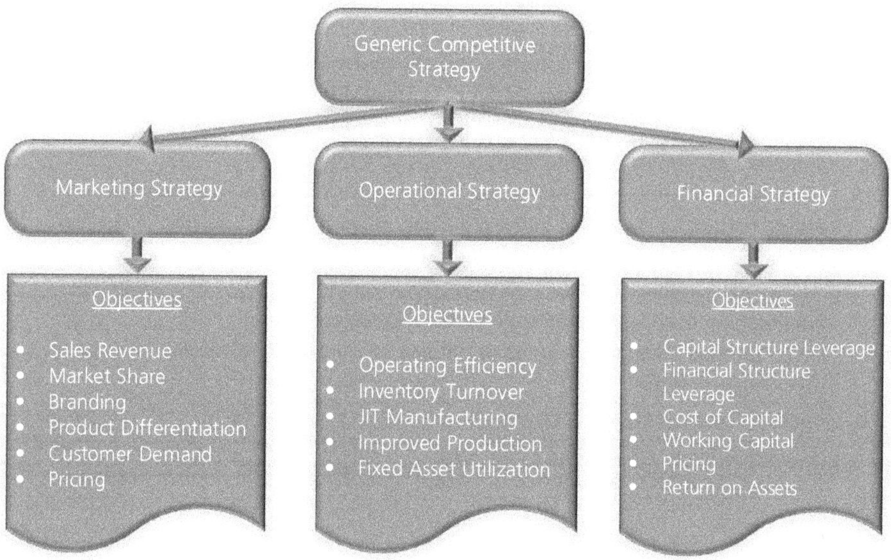

## REFERENCES

Bharadwaj A, El Sawy OA, Pavlou PA and Venkatraman N (2013) Digital business strategy: toward a next generation of insights. *MIS Quarterly* **37**(**2**): 471–482.

da Costa LS, Munhoz IP, Pereira L and Santos Akkari AC (2022) Assessing the digital maturity of micro and small enterprises: a focus on an emerging market. *Procedia Computer Science* **200**: 175–184.

Ghezzi A (2013) Revisiting business strategy under discontinuity. *Management Decision* **51**(7): 1326–1358.

Grooss OF, Presser M and Tambo T (2022) Balancing digital maturity and operational performance - progressing in a low-digital SME manufacturing setting. *Procedia Computer Science* **200**: 495–504.

Heshmatisafa S and Seppänen M (2023) Exploring API-driven business models: Lessons learned from amadeus's digital transformation. *Digital Business* **3**(**1**): 100055.

Meskendahl S (2010) The influence of business strategy on project portfolio management and its success – a conceptual framework. *International Journal of Project Management* **28**(**8**): 807–817.

Olson EM, Olson KM, Czaplewski AJ and Key TM (2021) Business strategy and the management of digital marketing. *Business Horizons* **64**(**2**): 285–293.

Perera S, Jin X, Das P, Gunasekara K and Samaratunga M (2023) A strategic framework for digital maturity of design and construction through a systematic review and application. *Journal of Industrial Information Integration* **31**: 100413.

Rafael LD, Jaione GE, Cristina L and Ibon SL (2020) An Industry 4.0 maturity model for machine tool companies. *Technological Forecasting and Social Change* **159**: 120203.

Reis J, Amorim M, Melão N and Matos P (2018) Digital transformation: a literature review and guidelines for future research. *Trends and Advances in Information Systems and Technologies* **1**(**6**): 411–421.

Sishi MN and Telukdarie A (2021) Supply chain energy sustainability with artificial intelligence. *Paper presented at the 2021 IEEE Technology & Engineering Management Conference-Europe* (TEMSCON-EUR), pp. 1–6.

Singh A, Klarner P and Hess T (2020) How do chief digital officers pursue digital transformation activities? the role of organization design parameters. *Long Range Planning* **53**(**3**): 101890.

Tutak M (2022) *Business Digital Maturity in Europe and Its Implication for Open Innovation.* MDPI AG, Basel. https://doi.org/10.3390/joitmc8010027

Vial G (2019) Understanding digital transformation: a review and a research agenda. *The Journal of Strategic Information Systems* **28**(**2**): 118–144.

Zhai H, Yang M and Chan KC (2022) Does digital transformation enhance a firm's performance? evidence from China. *Technology in Society* **68**: 101841.

Zezulka F, Marcon P, Vesely I and Sajdl O (2016) Industry 4.0 – an introduction to the phenomenon. *IFAC-PapersOnLine* **49**(**25**): 8–12.

Ziyadin S, Suieubayeva S and Utegenova A (2020) Digital transformation in business. *Paper Presented at the Digital Age: Chances, Challenges and Future* **7**: 408–415.

Arnesh Telukdarie
ISBN 978-1-83608-365-8
https://doi.org/10.1108/978-1-83608-362-720241010


# Chapter 8
# Digital water architecture

A Telukdarie and R Mongwe

## Introduction

A fourth industrial revolution (4IR), also known as Industry 4.0, is on the horizon and with it comes a wave of technological advancements that are expected to revolutionise every aspect of society. One area that will be impacted is water management, which faces numerous challenges. The fusion of physical, digital and biological systems will usher in an era of unprecedented innovation and interconnectedness. One of the critical challenges that humanity faces is the sustainable management of water resources. As the global population continues to grow, water scarcity is becoming an increasingly pressing issue. The 4IR provided an opportunity to revolutionise water management and ensure a sustainable future for all.

A team of dedicated scientists, engineers and innovators is working tirelessly to harness the power of the 4IR for water management at the University of Johannesburg, led by Professor Telukdarie, a renowned water resource expert. The team aims to develop an intelligent water management system that could optimise water usage, conserve resources and respond to changing environmental conditions. The team works hand in hand with Johannesburg Water, a city of Johannesburg municipal entity tasked with the provision of potable water to approximately 5.6 million residents of Johannesburg.

Several factors, including rapid urbanisation, a shifting climate, a growing population and an aging water distribution infrastructure, are creating new difficulties for the water industry in the coming years. The availability of water for people and businesses, as well as its quality, depends on overcoming these obstacles. 4IR solutions include the use of Internet of Things (IoT) sensors to collect data on water flow, water leakage, water balance and water quality (David *et al.*, 2022). IoT sensors can identify water-related issues, alerting decision-makers to potential issues (Daigger *et al.*, 2019). Technologies like advanced data analytics, cloud storage and machine learning are also on the rise. Water management is still a challenge in the twenty-first century for several reasons.

### Aging water infrastructure and maintenance of water assets

Due to the aging of equipment and related software tools in the water infrastructure, the water distribution network (WDN) constantly upgrades its equipment to keep the network operating. Additionally, water availability and uptime are reduced as the water infrastructure ages. Furthermore, the WDN needs constant maintenance. Thus, there are constant unplanned and planned water supply interruptions. Water utilities can introduce 4IR, including IoT and advanced automated processes that allow water supply, water treatment and infrastructure maintenance to mitigate some of the challenges related to the aging water infrastructure, which improves operations and maintenance through preventive maintenance.

## Management of water and electricity costs

A great deal of electricity is required to keep the water infrastructure operational daily, particularly the pumping system, which uses a considerable amount of power. Continual energy price changes pose a significant challenge for the water industry. To reduce the cost of energy, on-site renewable energy generation and storage assets are being introduced to water-hungry sites in Australia. Demand responses can be supported by renewable generation systems called distributed energy resources using Industry 4.0 technologies.

## Environmental and sustainability issues

Urbanisation, climate change, sustainable energy sources, flooding and sewer spillage are among the environmental challenges facing the water industry. Population growth in metropolitan areas can raise the risk of stormwater, which, if improperly managed, can contaminate streams, cause erosion and worsen flooding. Another major difficulty facing the water sector is its inability to control sewer obstructions, which eventually flow into the main network and pose serious health problems that could be expensive to clean up.

## Growth in urban population and rural services

Recent years have seen a clear increase in the population of urban areas, which presents a significant challenge for the water industry in terms of meeting rising water supply demand, maintaining water quality and reducing environmental risks brought on by urban sprawl. Using Industry 4.0 technologies, water supply management and customer service can be improved. Demand management systems can integrate asset or site data with external information data sets to enable real-time monitoring of water resources and consistent forecasting of issues relating to water.

## Aging workforce and shortage of skills

The current workforce within the water industry is used to the traditional method of managing water resources and infrastructure. In addition, it is predicted that many workers in the water sector will retire on average within the next ten to fifteen years. New digital technology skills will be needed in the workforce to successfully implement 'Water 4.0' technology as Industry 4.0 is introduced to the water sector. The lack of personnel who are aware of the implications of the new smart water technology is one of the biggest problems facing the water sector. Data analysis and data literacy, network security, interpersonal communication skills, problem-solving and critical thinking abilities are just a few of the specialised skills needed by the water industry workforce of the twenty-first century.

Until the advent of 4IR, the supervisory control and data acquisition (SCADA) and other operational technology systems used by water distribution networks were completely disconnected from the company's IT infrastructure. With the advent of the 4IR and its attendant innovations in areas like big data analytics (BDA), blockchain, simulation and modeling semantics, cloud computing, automation, cyber-physical systems (CPS), industrial internet of things (IIoT) and robotics, operational technology systems like SCADA must make their data available to other systems (e.g. IT systems), both inside and outside of the enterprise to be fully digitalised. These coordinated efforts will improve data-driven decision-making by bridging the gap between SCADA and its components' IT networks. Currently, many WDN operations are performed by hand, including leak detection, water quality monitoring and meter readings. Due to a declining unskilled labor population base and an increase in educational attainment, businesses are looking to digital transformation to reduce the need for labour-intensive manual processes and free up scarce human resources.

## Advantages of a digital WDN

The main advantages of a digital WDN are expected to be as follows.

### Provides customers with real-time water consumption information to help them make informed decisions

Using technologies such as automated meter readings and smart water-saving devices, a digital WDN will be able to provide customers with real-time feedback on their water and energy usage. Customers can now make educated decisions about water efficiency in their homes and businesses. In addition, automated meter readings will provide more precise demand predictions for optimising pumping schedules, water turnover in reservoirs and the volume of water that needs to be pumped and treated.

### Monitoring the condition of assets in real-time for preventive maintenance

Modern pipeline replacement projects can benefit from the use of risk-based models made possible by the application of advanced sensing technologies. This allows water utilities to better plan and schedule pipe replacements and rehabilitations, increasing the likelihood that the appropriate pipes will be replaced at the appropriate times (Li *et al.*, 2018).

### Monitoring of water pressure and water quality in real-time to improve planning and network operations

To reduce the risk of pipe bursts and maximise water conservation, water utilities can use real-time sensor and meter data to keep tabs on hydraulic and water quality conditions across their networks, allowing for rapid response to leaks and early detection of pipe stress. Water quality monitoring of distribution pipes is another useful method for foreseeing the onset of contamination. To prevent further flooding, damage, water loss or the spread of contaminated water, automated valves can respond by closing valves in the affected areas. Real-time sensor data can be used to replace the costly, time-consuming and laborious traditional calibration process in several applications, including the calibration of hydraulic models.

To monitor, manage and detect leakages, water utilities across the globe are adopting digital WDN. Johannesburg Water, also known as Joburg Water, is one of these water utilities. Joburg Water, a municipal entity of the city of Johannesburg, lost approximately 38% of its bulk water supply in 2019 due to damaged infrastructure. The city of Johannesburg receives 1.6 billion litres of drinkable water per day from Rand Water and distributes it through a WDN that includes 12 288 kilometres of pipe, 127 reservoirs and water towers, and 37 [illegible] pump stations. The wastewater is then collected, treated and [illegible] using 11 710 km of wastewater networks and 38 sewage pump stations. There are six wastewater treatment works (WWTW) operated by Johannesburg Water, one of which is a biogas-to-energy plant that converts methane gas into energy. The annual rate of water main failures per 100 kilometres of pipe was 221.16 in 2019, while the annual rate of clogged sewerage was 286.08 per 100 km in 2019. On average, the rate of infrastructure failure in water pipes has been rising over the past few years. The deteriorating condition of the pipe network was largely to blame. In mitigation of this, Joburg Water teamed up with Professor Telukdarie's team.

## Joburg water towards digital WDN

Various steps are being taken towards the digitisation of Joburg Water. The University of Johannesburg acts as an advisor and consultant for some of these steps. Other stakeholders, including SMEs and research centres, also play a major role in the transition. To date, two projects have been concluded, including the integration of small and medium enterprises (SMEs) for Industry 4.0 in the South African water services sector, and testing and proof of concept for automated leak detection using wireless sensors.

### Integration of small and medium enterprises for Industry 4.0 in the South African water services sector

Most economies' foundations rest upon the success of their SMEs. To stay competitive in the market, SMEs must continually adjust their products and methods of production. The digital concept, especially Industry 4.0, creates new innovative business opportunities and increases SME productivity. Only a few businesses, however, are prepared to adapt to ongoing transformation, particularly to a new wave of digital technologies that brings both challenges and opportunities for SMEs. Through the expectations of the University of Johannesburg, two SMEs were identified. The SMEs ensure the integration of SMEs in Industry 4.0 for the municipal water sector within a South African context.

The first SME focuses on innovative, cutting-edge solutions for the water services industry. Using the Kingpin Aqueduct automated leak detection system, the company oversees smart leak detection, monitoring, leak repair and maintenance. This includes cloud-based leak monitoring software and hardware for daily analysis of acoustic signals emitted by fixed sensors. The signals trace and pinpoint small to large leaks while preventing pipe bursts. This system provides real-time information on leaks as soon as they begin to develop, pinpointing their exact location long before any damage is done. The system is predicated on interconnected sensors that use 3 G cellular communication, global positioning system (GPS) synchronisation and automatic signal processing in the cloud. A small, designated area will be tested to reduce the amount of water that is lost to evaporation. The proposed network would collect both historical and real-time data, from which a real-time configuration could be derived. This involves reviewing the mobility platform with a subject matter expert, which fills in the gap between data collection in the field, data analysis and decision-making.

The second SME is an innovative, technology-driven civil engineering consultancy that works with clients across a wide range of industries. Using criteria established by the International Water Association (IWA), the company manages water supplies and performs leak detection, WC/WDM and water balancing. By identifying novel systems that are interoperable with Johannesburg Water's existing water management systems like Integrated Quality Management System (IQMS), Geographic Information System (GIS) and Supervisory

Control and Data Acquisition (SCADA) and by proposing an automated water balancing system that can be implemented as part of the CPS for the water utility, the company will provide civil engineering technical support. By acquiring all the necessary data, the SME plays a vital role in ensuring a feasible water-balancing system is implemented. Among the responsibilities of the SME are water balance, network sectorisation, smart bulk meter audit, pressure and zone management, civil engineering technical support, Water Conservation and Water Demand Management (WC/WDM) model analysis and capacity building. The two SMEs work together with Joburg Water and the University of Johannesburg towards the digitalisation of Joburg Water's WDN.

## Testing and proof of concept for automated leak detection using wireless sensors

The team's first task was digitising water demand management in pursuit of reducing high levels of nonrevenue water. The first step in this ambitious project was conducting a pilot test. The results of the pilot project will help Johannesburg Water Company create a comprehensive plan for implementing digital water demand management. In particular, the pilot program intends to implement digitalisation in the areas of leak detection, advanced pressure management, advanced metering, repairs and maintenance, customer relations management (CRM) and water audit or balance, all of which will be defined in a business case or strategy. Through the Process, Energy and Environmental Technology Station (PEETS), two SMEs, Boffin & Fundi (B&F) Smart Innovative Solutions and Epitome Consulting were enlisted to conduct pilot testing and proof of concept on the various available innovative technologies for water demand management. These sensors were dispersed throughout the water supply system, from reservoirs and treatment facilities to pipelines and homes. They generated a massive amount of real-time data by continuously monitoring things like water quality, usage patterns and infrastructure performance.

Country View, in the Midrand region, was selected by Joburg Water and B&F as the best location for testing the technology. There was successful deployment and operation of both stationary and portable units. The Joburg Water automated leak detection monitoring system now includes GIS data that was requested from the Infrastructure Planning Department. Since GIS data showed that the piping infrastructure was made up of asbestos cement pipes, shorter distances of 200 m to 250 m were provided for the embedded wireless sensors. The crew ran into issues at Country View North because there were not enough installation points to accommodate the necessary distance and variety of sensors. In some cases, over 250 metres separated the installation sites. Contrary to the GIS data presented, the installation points shown on the system represent sewer valves.

After being installed on-site, sensors were set to automatically record and transmit data at 2:30, 3:30 and 4:30 every morning. Except for one sensor, which moved the following week, all others immediately began receiving a strong cellular signal. Throughout the pilot test, the sensors' health was tracked remotely; one alert was generated due to a low signal from a sensor attached to an air valve. Later, the sensor was moved to a nearby fire hydrant.

The site investigation for the automated leak detection pilot in Country View confirmed water leakages, lending credence to the findings that the sensors detected water leaks and generated an alert. These alerts are fully automated and each has a unique alert identification that can be used to create a work order for the repair and maintenance teams that will visit the site and make the necessary repairs. The precise GPS coordinates were incorporated into a work order.

The system found and analysed two leaks in Sneeugrass, which is consistent with the sample data; however, because of the interplay between the two locations, the intensity of the leaks changed on different occasions about a hundred metres from sensor 1, where the leakage was less severe. Further, this data enables intelligent decision-making. Professor Telukdarie and his team leveraged the power of artificial intelligence (AI). They plan to develop sophisticated algorithms that analyse data streams, identifying patterns, trends and anomalies. This AI-driven system could predict water demand, detect leaks and inefficiencies, and optimise the distribution of water resources. This will benefit both Johannesburg Water and water consumers.

## Conclusion

One of the critical challenges that humanity faces is the sustainable management of water resources. As the global population continues to grow, water scarcity became an increasingly pressing issue. Several factors, including rapid urbanisation, a shifting climate, a growing population and an aging water distribution infrastructure, are creating new difficulties for the water industry in the coming years.

The University of Johannesburg's team, scientists and engineers aim to develop an intelligent water management system to optimise water usage, conserve resources and respond to changing environmental conditions.

## Teaching notes

### Synopsis

Water is the fundamental element of life; it is also a working fluid for some of the major chemical processes, such as thermal energy generation. Additionally, water is used for manufacturing and domestic use. However, it is scarce, and water utilities such as Johannesburg Water have challenges distributing the water. Water distribution and management challenges in South Africa are well documented. Water losses exceeding 40%, coupled with limited resources on the supply side, result in financial losses and negative environmental impacts. New technologies introduced by 4IR can mitigate some of the challenges faced by water utilities. However, implementation challenges exist. Johannesburg Water decided that to benefit from the new era of 4IR, it needed to work hand in hand with 4IR experts, which resulted in the formation of a research and consulting partnership between the University of Johannesburg and Johannesburg Water. The approach is to develop a detailed landscape and align it with digital architecture. Use the case study entity to gather and test the comprehensiveness of the single architecture.

### Teaching objectives

The case is intended for students at the Johannesburg Business School (JBS) at the Master of Business Administration (MBA) level. The focus of the case is to understand and explore the challenges faced by water utilities. The case will have guidelines for how water utilities can work with SMEs to mitigate some of the problems they face. The following teaching objectives were formulated.

- **To explore the challenges faced by water utilities:** The case study highlights the challenges faced by water utilities worldwide and how digital technologies can be used to mitigate some of these challenges.
- **To understand and explore the benefits of 4IR in the water industry:** Students will learn about 4IR technologies that are being used in the water industry to solve various challenges in water distribution. Students will also learn more about the three main benefits of these technologies.
- **To explore the contribution of SMEs in the South African water sector:** Students will learn about how SMEs helped Joburg Water with the digitalisation of the water distribution network. Exploring the role of SMEs in the South African water sector is essential for unlocking potential contributions to innovation, local development and sustainable water management.

## Pedagogy

Various concepts are addressed in this case, of which the most important are challenges faced by water utilities, the adoption of 4IR to mitigate the challenges, the benefits of 4IR in the water sector and public-private sector cooperation.

There is no right or wrong answer when it comes to solving the case's problems. Students should examine each concept addressed in the case and identify the solutions used by the main character.

Students can apply the solutions from this case to any type and any size of organisation, resulting in the long-term viability of a sound business.

## Research methods

Specifically, the objective of this project is to develop and implement a digitalised blueprint for the city of Johannesburg's water demand, which will focus on advanced pressure management, smart meters, automatic leak detection and the reduction of nonrevenue water consumption. Water demand management will be benchmarked for the fourth industrial revolution. Through the inclusion of small and SMEs during the implementation of Industry 4.0 for the city, this project promotes the University of Johannesburg (UJ), Process, Energy and Environment Technology Station (PEETS) and Johannesburg Water interventions concerning the fourth industrial revolution vision. It is important to identify a pilot water supply zone and deploy wireless sensors to prove the concept of implementing a leak detection and monitoring system to successfully implement automated water demand management.

## Keywords and subject code

AI, Water Challenges, Digital Water Distribution Network, SMEs, 4IR and IoT

## Assignment questions

- What are the challenges faced by water utilities?
- What are the 4IR technologies that can be used in the water sector?
- What are the benefits of 4IR in the water sector?
- What is the importance of SMEs in the water sector?

## Suggested readings

Nthutang P and Telukdarie A (2018) Integration of small and medium enterprises for Industry 4.0 in the South African Water Services Sector: A case study for Johannesburg Water. *2018 IEEE International Conference on Industrial Engineering and Engineering Management (IEEM)* (pp. 1206–1210). IEEE.

Nthutang P, Telukdarie A, Medoh C and van Rensburg NJ (2019) Testing and proof of concept for automated leak detection using wireless sensors: a pilot study for Johannesburg Water Company. *2019 IEEE International Conference on Industrial Engineering and Engineering Management (IEEM)* (pp. 479–483). IEEE.

## Teaching plan and timing (time should be at least 90 minutes)

- *15 min*: Class introduction: The instructor should distribute the case as prereading. If offered in South Africa, one of the stakeholders (UJ, Joburg Water, or the two SMEs) should come to class and do the introduction from the heart.
- *30 min*: Assignment question 1: The instructor should get the students to work in pairs discussing the challenges faced by water utilities, the adoption of 4IR to mitigate the challenges, the benefits of 4IR in the water sector and public-private sector cooperation.
- *15 min*: Assignment question 2: The facilitator should request the students to identify three recommendations related to the challenges faced by water utilities in the Fourth Industrial Revolution (4IR). The instructor should explain that each group should discuss and come up with three practical recommendations.
- *15 min:* Assignment question 3: The facilitator should request one group to come and present the findings or recommendations related to the water utilities. The instructor should explain that each pair must present the benefits of the solution and how it is a 4IR solution.
- *10 minutes*: Assignment question 4: The facilitator should request that the students identify the importance of SMEs in the water sector.
- *5 min*: Conclusion: The instructor should wrap up the lesson by reflecting on the major challenges that the students identified and their recommendations.
- As it is a story being told, this case should be handled through discussion rather than PowerPoint presentations.

## Analysis of assignment questions

For each assignment question, a sample response of roughly 200 words should be provided. The application of models, frameworks and concepts from textbooks and journal articles should be included in the response to deepen the analysis and demonstrate an understanding of current academic thinking in the area of study. The recommended core readings listed above should be used along with other resources.

## Key learnings

The instructor should wrap up the lesson by reflecting on the major challenges that the students identified and that they could not identify. Furthermore, the instructors should provide the lessons with the solutions that they believe are appropriate for assignment questions as a form of conclusion.

REFERENCES

Coward T, Tribe H and Harvey AP (2018) Opportunities for process intensification in the UK water industry: a review. *Journal of Water Process Engineering* **21**: 116–126.

Daigger GT, Voutchkov N, Lall U and Sarni W (2019) *The Future of Water: A collection of essays on 'disruptive' technologies that may transform the water sector in the next 10 years*. Washington, D.C: Inter-American Development Bank.

David LO, Nwulu NI, Aigbavboa CO and Adepoju OO (2022) Integrating fourth industrial revolution (4IR) technologies into the water, energy & food nexus for sustainable security: A bibliometric analysis. *Journal of Cleaner Production* **363**: 132522.

Li J, Lee S, Shin S and Burian S (2018) Using a Micro-Test-Bed Water Network to Investigate Smart Meter Data Connections to Hydraulic Models. *World Environmental and Water Resources Congress 2018: Hydraulics and Waterways, Water Distribution Systems Analysis and Smart Water*. American Society of Civil Engineers, Reston, VA, USA.

Arnesh Telukdarie
ISBN 978-1-83608-365-8
https://doi.org/10.1108/978-1-83608-362-720241011


# Chapter 9
# Reproducible and sustainable cyber-physical system (CPS) implementation in multinational enterprises

A Telukdarie and E Buhulaiga

## Introduction

'In line with the Saudi Vision 2030 and to create a vibrant society with a thriving economy, our objective was not only to create a successful standalone cyber-physical system (CPS) but to create a reproducible and sustainable CPS-as-a-Service to Saudi Arabia and the world! (Figure 9.1). Reproducible and Sustainable CPS-as-a-Service)', said Dr. Eyad Buhulaiga, the CPS Project Manager, at the closure meeting of the successful CPS project. Eyad Buhulaiga, Operational Technology Project Manager, was requested to lead the implementation of manufacturing operations management (MOM) in a greenfield refinery within a multi-national enterprise (MNE) energy giant.

The refinery is part of a process complex containing a refinery, terminal and power plant. The scope of MOM was limited to the implementation within the refinery only and included the integration of the MOMs for the terminal and power plant. The scope of the three MOMs had been defined and assigned to separate engineering, procurement and construction (EPC) contractors, and the interface points between the three projects had already been identified and scoped. Integration and interfacing between the three projects and the three EBCs pose challenges resulting from the need to align the timeline and deliverables between the three projects and set priorities to minimise delays caused by project dependencies. Eyad knew that managing the project would be a nightmare and he started questioning whether he should continue with the project as is. 'Would there be a better way to align the three projects and optimize the scope?', Eyad speculated.

## Multinational enterprise (MNE) energy giant

The MNE energy giant has embarked on a digital transformation journey with a mission of being an integrated energy company and an objective of achieving operational excellence (OE) to improve efficiency, reliability and profitability, as well as the health, safety and environment (HSE) targets. The journey includes the deployment of several Industry 4.0 digital solutions that help to bridge the automation gap in the four OE pillars. These digital solutions complement the existing manufacturing execution systems (MES) and MOM implementations.

## IR3.0 MES and MOM

As the MNE energy giants have hundreds of operating facilities scattered around the world, their reliance was mainly third industrial revolution technologies, with a focus on MES implementation. For the past two decades, the MNE has consistently adopted new technologies to help maintain competitiveness and spearhead the industry, while conforming to the four OE pillars. The challenge that MES imposes is that it is designed to fit standalone implementations that do not expand to multisite and multinational implementations. MES conforms to specific requirements that are customised to meet the manufacturer's requirements. This was why the MNE changed the scope to MOM, to allow for extending the implementation to more than one site. However, due to the slow adoption of cloud technology in the energy industry and the lack of a standard methodology of implementation that can be followed to extend across boundaries in different geographical locations, the design of the three new operating facilities included three different MOM projects!

## ISA-95 (Control-Enterprise integration) standards

The MNE adopted ISA-95 as a standard to follow for the MOM projects. ISA-95, based on the Purdue model, divides automation layers in any manufacturer into five different layers (i.e. Layer 0: Asset Controllers, Layer 1: Sensors and Actuators, Layer 2: PCS, Layer 3: MOM, and Layer 4: ERP) (Figure 9.2: MOM implementation based on the Purdue computer integrated manufacturing (CIM) Model). Coming from an IT and software engineering background and being a service-oriented architecture (SOA) and cloud computing practitioner, Eyad, wearing the MNE hat, thought of optimising the scope to minimise the implementation complexity of capital expenditure (CAPEX) and operational expenditure (OPEX). Eyad was lucky to have a general manager, assigned as a thought leader, heading the greenfield refinery complex, where the three project teams reported to him. Eyad discussed the consolidation idea with the general manager, and the idea was well received. Eyad was fully empowered, and he was advised to 'ask for forgiveness and not ask for permission', if the ultimate objective is met!

## Reproducible CPS to the rescue

Eyad, with the support of his mentors from academia and industry, as well as his innovative and young team members, has researched the best way of implementing MOM in a reproducible manner. This is where he found that CPS is the right approach to follow. CPS extends MOM and MES implementation in a systematic architecture called the 5Cs architecture that allows manufacturers to.

- *Connect* with physical systems (physical space).
- Fetch and *convert* data into contextualised and usable format.
- Provide a *cyber* and information system hub that integrates all information from all connected data sources and creates a cyberspace.
- Make use of big data by deploying advanced analytics and *cognitive computing*.

*Configure* role-based dashboards.To enhance it even further and to cater to reproducibility, Eyad added a 6th 'C' to make it *cloud-based*, which covers all aspects required by MNE to reap the benefits of CPS-as-a-Service that include reconfigurability, agility, functionality, performance, interfaces, sustainability and operational and economic benefits.

## Business case and approval to proceed with consolidated CPS implementation

Eyad built a business case with an optimised scope to secure the general manager's approval. The business case included quantitative and qualitative benefits to the greenfield complex. It provided an extended vision that allows for the implementation and investment to expand to cover not only the complex but also to extend beyond the country and MNE by commercialising and productising the implementation to allow for other energy firms to subscribe to the solutions.

## CPS-as-a-Service quantitative benefits

Eyad was always looking for ways to improve profitability and had read many studies that showed how implementing CPS could significantly improve profitability by 0.5–1.5 $/bbl of oil equivalent, depending on the current state of maturity and several business functions undergoing such transformation.

One day, Eyad decided to conduct his study, in cooperation with a management consulting firm, on the benefits of CPS-as-a-Service and the value of adopting CPS in a typical plant. He divided the benefits into several categories: capital efficiency, value stream optimisation, asset optimisation, process optimisation and environmental management. Each category had its own set of benefits that could be quantified. For capital efficiency, the team found supporting evidence that implementing CPS could reduce asset lifecycle cost by 15–20% and spare part inventory by 2–5%. In value stream optimisation, working capital could be reduced by 15–20% and raw material inventory by 5–15%. Asset optimisation could increase asset utilisation by 3–10%, save energy by 5–15% and reduce maintenance costs by 10–20%. Process optimisation could increase production yield by 3–10% and reduce quality losses by 5–10%. Finally, environmental management could reduce waste by 5–10%, which would also reduce corporate liability.

The reusability of infrastructure, pre-engineered solutions, reduction in project duration (resulting from doing the project only once for the three sites), hosting infrastructure, software licencing, data protection, data centres, manpower involved, energy charges, communication and other operating costs are all parameters contributing to the calculated benefits of building reproducible CPS as opposed to stand-alone MOM implementation. To demonstrate the value, the project manager used the three MOMs implementation as the basis for calculation and comparison between standalone implementations as opposed to CPS-as-a-Service. The extension of CPS results in a major cost reduction of more than 50% for CAPEX (Figure 9.3. CAPEX reduction) and 57% for OPEX (Figure 9.4: OPEX reduction). The details of the findings are available in the addendum.

Eyad was excited by these findings and used them as a justification for the merger of the three projects. Management and financial controllers were impressed by the potential benefits of CPS and approved the merger.

## CPS-as-a-Service and centre of excellence

Eyad was challenged by the refining management, as they could not understand the refining context in the business case or how that implementation would be sustained, given the skillsets and aging levels of the current workforce of the complex's employees. As is his habit, with Eyad's refusal to accept 'no' for an answer, he collaborated with subject matter experts and illustrated how the fourth industrial revolution (4IR) and CPS are capitalising on the data explosion, availability of instrumentation and data acquisition systems (Table 9.2). This allows for the implementation of digital transformation technologies within the manufacturing industry.

Eyad demonstrated how CPS is providing a digital representation of the plants (digital twin) by way of the 5Cs architecture that allows manufacturers to *(i) connect* with assets (e.g. heater), done by providing real-time monitoring to the heater parameters (e.g. coil outlet temperature (COT), control valve opening, lab tests and inspection rounds test), *(ii) convert* data into contextualised and usable format (i.e. provide the heater's data such as the real-time reading of COT within the acceptable ranges to allow for faster decision making), *(iii)* provide *cyber* hub that integrates all the information from all connected data sources, *(iv)* make use of the big data by deploying advanced analytics and *cognitive* computing to evaluate root causes of the fouling to provide corrective actions and *(v) configure* role-based dashboards (e.g. operator, consultant and supervisor). After that, the project manager demonstrated how adding the sixth 'C' (*cloud-based* implementation) to include the power of the cloud to the CPS 5Cs and creating the CPS 6Cs model, as demonstrated below, will add value to the implementation.

Aligning with the thought leadership of the general manager, Eyad also stressed institutionalising the use of CPS, implementing a complete change management program as part of the project and creating a Center of Excellence (CoE) to monitor the performance of the global locations. It also allows for tapping into the expertise of process licensers, OEMs and third-party consultants, as deemed necessary. Figure 9.6: demonstrates the CPS-as-a-Service 6Cs framework (Figure 9.6: Proposed CPS 6Cs). This has demonstrated how CPS could help MNE foster collaboration among multidisciplinary teams globally and move from standalone CPS implementation into CPS-as-a-Service for multinational businesses.

## Sustainability of CPS-as-a-Service and return on investment (ROI) calculation

The general manager was convinced of the benefits and wanted to support the case by adding the impact of the consolidation on sustainability in addition to the tangible and intangible benefits of CPS-as-a-Service (Table 9.1). Centralisation of CPS has a positive sustainability impact by reducing $CO_2$ emissions and power consumption resulting from reusability. The project manager and their team used the available information from the Aviation Environment Federation, stating that $CO_2$ emissions per person on an annual basis is 20 tonnes. By reducing the number of personnel involved in developing and managing CPS (e.g. travelling, hosting and using air conditioning), there is a great reduction in $CO_2$ emissions (up to five times less). The comparison between standalone against CPS-as-a-Service implementation from power consumption and emission perspectives is demonstrated in the addendum (Figure 9.5: Reduction of $CO_2$ emission and power consumption resulted from CPS-as-a-Service).

As detailed in Table 9.3, the total energy consumption and maintenance optimisation resulting from deploying CPS is the same for both types of implementations, as the operating facilities realise the value of the 4IR system and improve reliability, efficiency, effectiveness and, hence, profitability. The CAPEX, OPEX, NPV and ROI differ, although to the advantage of the CPS-as-a-Service. This is due to the increase in upfront CAPEX needed to establish standalone CPS as well as the operating costs that are paid fully by a single operating facility as opposed to being distributed among the MNE's operating facilities. This resulted in a positive NPV and an ROI for the CPS-as-a-Service that justified the investment in the deployment of the 4IR. The following diagram displays the ROI with the payback period and compares the ROI of CPS-as-a-Service against standalone CPS, demonstrating that CPS-as-a-Service payback and ROI start in the third year after implementation (Figure 9.7: CPS ROI comparison (CPS-as-a-Service against standalone)).

## Postmortem CPS-as-a-Service analysis

The implementation of the CPS-as-a-Service was successful and the MNE conducted a postmortem analysis that validated the assumption that it was positioned to approve the consolidation case. It established the nucleus of an entirely new revenue stream for the MNE as the implementation has also been productised to provide the CPS-as-a-Service to other MNE operating facilities across different locations and boundaries, with the possibility to commercialise it for other energy companies. This supports the company's vision of being a digital entrepreneur and gaining more financial growth from nontraditional core businesses.

Eyad's success story and thought leadership spread throughout the industry and many other companies began to adopt CPS as well. Eyad became a respected expert in the field and continued to push the boundaries of what was possible with CPS. He was proud to have made a positive impact on the industry and to have helped many companies improve their profitability.

## Exhibits

Figure 9.1 Reproducible and sustainable CPS-as-a-Service (author's own)

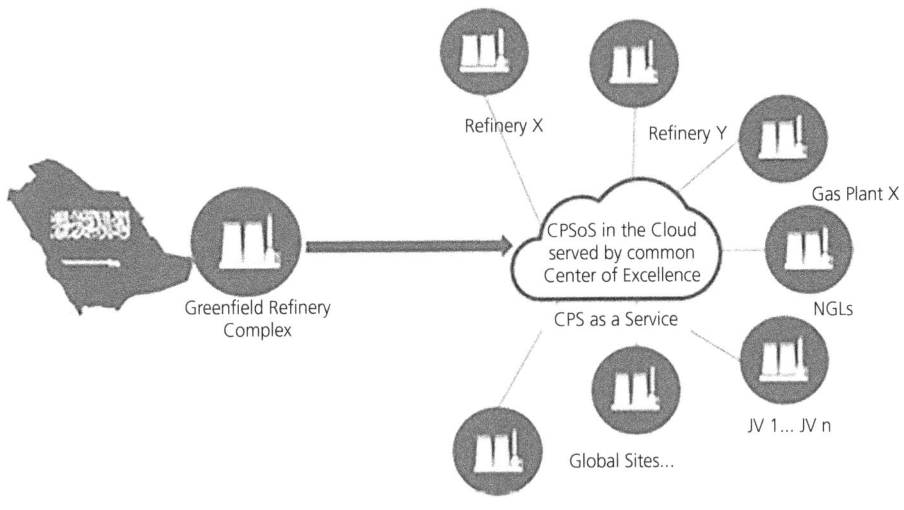

Figure 9.2 MOM implementation based on purdue CIM model (author's own)

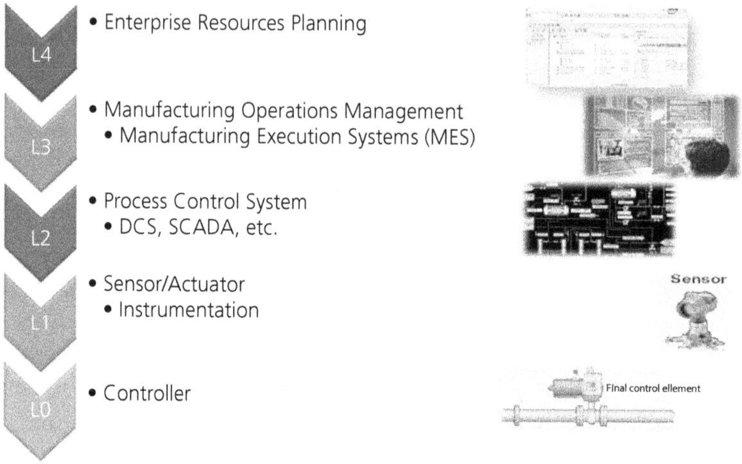


Figure 9.3 CAPEX reduction (author's own)

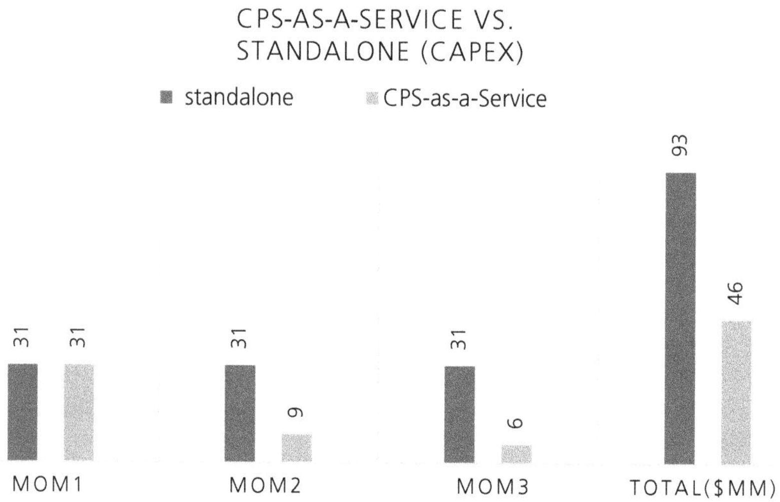

Figure 9.4 OPEX reduction (author's own)

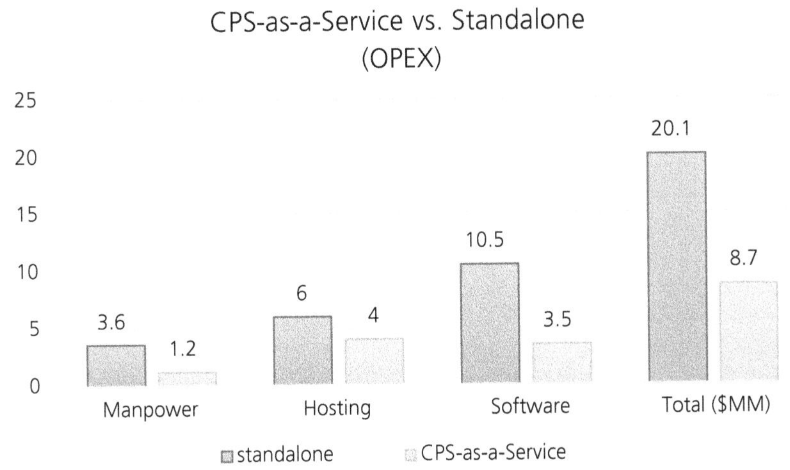


Figure 9.5 Reduction of $CO_2$ emission and power consumption resulted from CPS-as-a-Service (author's own)

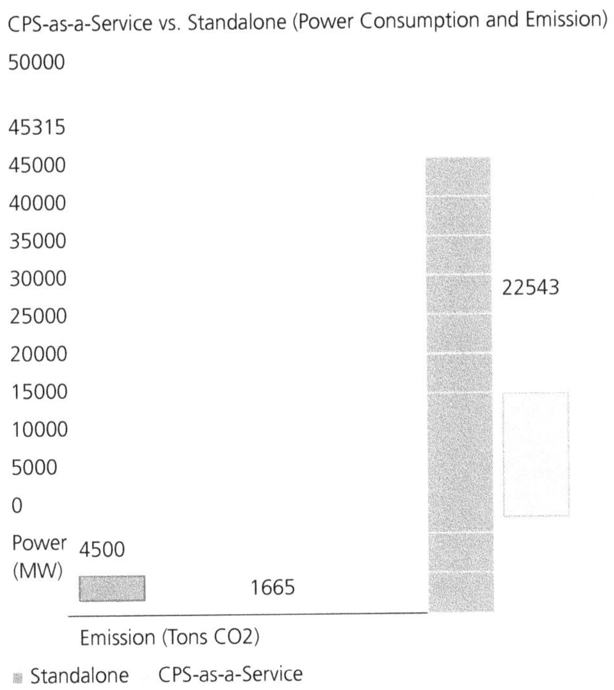

Figure 9.6 Proposed CPS 6Cs (author's own)

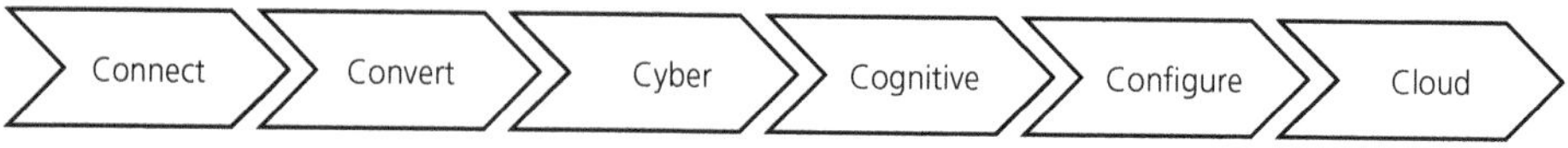


Figure 9.7 CPS ROI comparison (CPS-as-a-Service against standalone) (author's own)

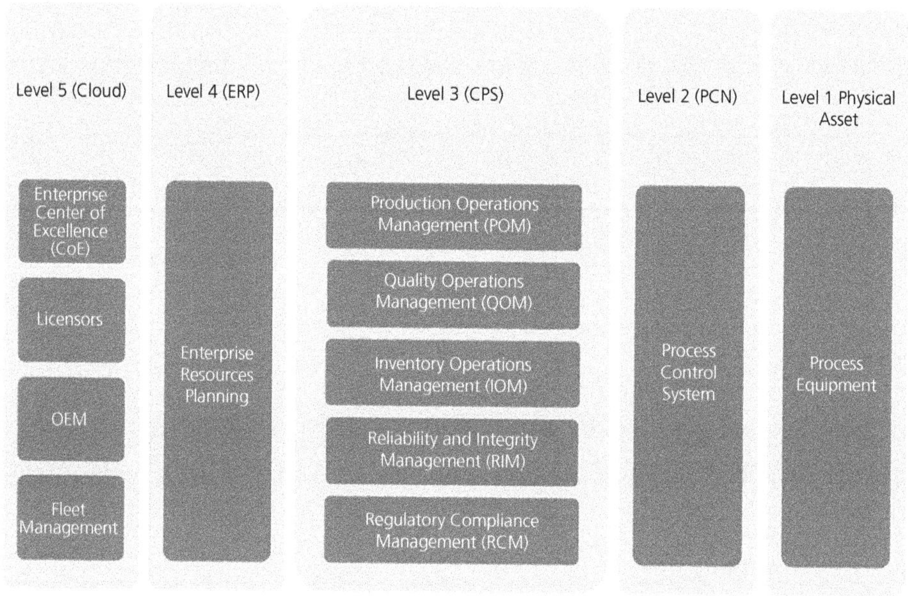


Table 9.1 Effort needed for standalone implementation

| Item | MOM Project 1 | MOM Project 2 |
|---|---|---|
| Setup the Datacentre | Need to be established | Need to be established |
| Procure Software License | Need to be established | Need to be established |
| Establishing integration with ERP | Need to be established | Need to be established |
| Establishing integration adaptors and configuring the role-based dashboard | Need to be established | Need to be established |
| Establishing and configuring the Integration Middleware System for intra-CPS systems | Need to be established | Need to be established |
| Building major asset templates | Need to be established | Need to be established |
| Establishing a network and data centre to host the servers | Need to be established | Need to be established |

Table 9.2 Summary of calculation parameters

| Item | CPS Project 1 CPS Project 2 |
|---|---|
| Set up the Datacentre | Infrastructure-as-a Service |
| Procure Software License | Software-as-a-Service and Platform-as-a-Service |
| Establishing integration with ERP | Integration-as-a-Service |
| Establishing integration with Legacy Systems | Integration-as-a-Service |
| Establishing integration adaptors and configuring the role-based dashboard | Done once |
| Establishing and configuring the Integration Middleware System for intra-CPS systems | Done once |
| Building major asset templates | Done once |
| Establish a network and data centre to host the servers | Infrastructure-as-a Service |

Table 9.3 Effort needed for CPS-as-a-Service implementation

| Calculation Parameter | Standalone CPS | CPS-as-a-Service |
|---|---|---|
| Total Energy Consumption over the years ($MM) | 40.71 (same benefits for both types of implementations) | |
| Total Maintenance Optimisation over the years ($MM) | 81.42 (same benefits for both types of implementations) | |
| CAPEX/OPEX ($MM) | 256.65 | 116.84 |
| ROI (%) | −52% | 5% |
| NPV (3% discount rate) | −130.85 | 0.75 |
| Payback period (Payback period = The value of the year in which last negative cumulative cash flow occurred + (value of the cumulative cash flow in that year divided by the cash inflow in the next year)) | None | 7.00 |

# Teaching notes

## Synopsis

The cyber-physical system (CPS) implementation case is the story of a multinational enterprise (MNE) energy giant that initiated and developed a CPS initially developed a CPS for one facility and later expanded to encompass other operating facilities, across the world. The business is reproducible, sustainable and well-managed by the next generation of the MNE's task force. This case depicts a practical implementation supported by a strong low total cost of ownership, capital and operating cost to the MNE business. Understanding the different stakeholders within the value chain, the external environment and its challenges made MNE successful in that implementation led to more innovation, growth and commercialisation of the CPS the MNE and others, which converted the CPS from a cost element to a profit generation business. The case study originated as a modest small project that was designed with an end in mind: to evolve into a program and subsequently develop into a successful commercial enterprise in Saudi Arabia, offering valuable services to energy companies, ultimately achieving remarkable success. This implies that the business is sustainable and sound and will still exist decades into the future.

## Teaching objectives

- To interpret the traits of the MNE whose story is told in the case.
- To assess the role that reproducibility and sustainability played in the leadership approach of the MNE team in the story.
- To appraise the innovation, thought leadership and management skills portrayed in this case.
- To analyse the importance of understanding your environment and context when making business decisions.
- To value the recognition of business and financial principles in making decisions on the diversification and growth of businesses.
- To understand what makes a business sustainable and existing for many years to come.

## Pedagogy

Various concepts are addressed in this case, of which the most important are strategic thought leadership, converting challenges to opportunities, business skills and approaches, understanding the external environment, entrepreneurship, reproducibility and sustainability, leadership and management, diversification, growth, financial and investment planning, utilising opportunities, and change management principles.

Solving the problems related to the case is not related to a right or wrong answer. Students should take each concept addressed in the case and identify the solutions that the MNE applied in the case.

Students can generalise the solutions from this case to any business type and any size of an organisation, leading to the sustainability of a sound business.

## Case analysis

Based on a recent implementation at one of the energy giants, this case study provides insights into the conceptualisation, design, implementation planning, systems selections and change management of a full implementation at a greenfield refinery. The case study is detailed and includes the integration of the full control system into enterprise resource planning integration. It provides a standard framework for a reproducible CPS and discusses the key aspect of the design standardisation, while the ability to replicate the system across the MNE's global refineries is incorporated into the design. The case study also highlights how standardisation optimises the total cost of ownership from capital and operational expenditure (CAPEX/OPEX) and the impact on sustainability targets.

## Assignment questions

- Identify and analyse the characteristics of a successful CPS implementation.
- What makes a business sustainable over a long period?
- What role do reproducibility and sustainability play in business and financial decisions?
- Describe the leadership and management characteristics of MNE and what impact they had on the business's success.
- Describe how taking the external environment, bigger picture, change management and context into account played a role in the success of the MNE implementation and commercialisation.

## Teaching plan

- Class introduction: The case can be distributed to the students as prereading by highlighting the challenges MNE is facing during Industry 4.0 implementation. Maybe (if offered in the Kingdom of Saudi Arabia) one of the MNE task forces can come to class and do the introduction from the heart.
- Major topics: The following topics can be discussed and unpacked, linked to the questions – environment, entrepreneurship, reproducibility and sustainability, leadership and management, diversification, growth, financial and investment planning, utilising opportunities and change management principles.
- Supplemental teaching components: Add additional reading material or videos on the history of MNE.
- This case should be dealt with through discussion, not through PowerPoint presentations, as it is a story being told.

## Conclusion

This story of a very successful MNE who built and grew a reproducible CPS within a very challenging economic environment provides a wealth of information on environment, entrepreneurship, reproducibility and sustainability, leadership and management, diversification, growth, financial and investment planning, utilising opportunities and change management principles.

Students can learn from this real-life example and apply all these principles to their businesses today, regardless of the business type, size, or industry.

## Guidelines on writing the actual case study

- Develop the teaching objectives (after discussing this case, students will be able to understand the different ways of developing CPS in MNE).
- Identify the situation (developing a reproducible CPS in energy MNE, where MNE has multi-sites, most cost-effectively and sustainably).
- Identify the protagonist (seek approval to use the real name and title of the person; makes it much more 'real').
- Outline the situation from the protagonist's point of view.

- Identify the time frame and explain it in the first paragraphs.
- State the case problem very clearly.
- Outline > first draft > research (including interviews) > second draft (stakeholder approvals).

Arnesh Telukdarie
ISBN 978-1-83608-365-8
https://doi.org/10.1108/978-1-83608-362-720241012


# Chapter 10
# Digital systems analysis of the food and beverage (FoodBev) architecture in South Africa

A Telukdarie and L Makoni

## The transformative journey of the FoodBev Sector Education and Training Authority (SETA)

The FoodBev sector has various functions, processes and systems. FoodBev's enterprise processes, which constitute the core functional departments, are identified and mapped on an activity basis predicated on current practice as stipulated by the applicable standard operating procedures (where they exist). These processes represent the organisational baseline and the foundation upon which value improvement studies are conducted. The existing processes (as-is processes) are subject to various analyses resulting in the identification of value-adding prospects, opportunities for creating synergies through redundancy obliteration and optimised processes.

In a place where technological disruption has become the norm, the FoodBev SETA has demonstrated an acute awareness of the potential inherent in fourth industrial revolution (4IR) technologies. Beyond perceiving these technologies as mere tools, the organisation envisions them as catalysts for revolutionising its core operations. The quest for efficiency, effectiveness and heightened impact serves as the driving force behind SETA's resolute commitment to embracing the digital age. While the realm of 4IR integration is marked by various documented approaches, FoodBev SETA's journey introduces a distinct challenge: the exploration of a comprehensive strategy that unlocks the realm of best practices.

The FoodBev SETA examined the dichotomy of challenges and opportunities that could arise when they embarked on a quest to elevate their practices from good to exceptional. At the heart of FoodBev SETA's journey lies a fundamental shift towards a holistic approach to digital transformation. Beyond the mere adoption of advanced technologies, the organisation focuses on the development of an integrated system that harmonises processes, data flow and communication. This transformation is not confined to the realm of technology alone; it encompasses a cultural shift, necessitating a profound change in mindset and a commitment to embracing innovation at every level. It looks at the progression of FoodBev in digital maturity, the juxtaposition of the as-is and the to-be processes, and the overall systems analysis.

## Digital business transformation in the service industry: a comprehensive systems analysis

The business operates in different kinds of areas and environments, such as retail, manufacturing, services and banking. The most valuable characteristic of any business is its ability to adapt to the dynamic environment in which it operates. Digital business transformation is increasingly becoming a key strategy for many organisations as they are constantly challenged to improve their business processes and capabilities. It is crucial to redesign business functions incorporating digital technologies and to ensure that digital processes are put in place to increase business efficiency (Ghosh *et al.*, 2019). Research has shown that organisations that have a comprehensive digital system synchronised across all business functions are not only performing better, but they are also making a more meaningful contribution to both the domestic and global value chains. Such is the goal of the FoodBev SETA, given that the hospitality industry in South Africa is one of the most performing sectors in Africa.

Digitalisation requires various strategic and operational knowledge and analysis. The FoodBev SETA recently realised that a constant mechanism and solutions for systems and maturity are needed for the sector to remain efficient and competitive. SETA emphasised that what is needed is knowledge of integrated architecture, systems and business processes. The ability to analyse and design an integrated business is a challenge for many FoodBev enterprises in the country. This is because a business modelling process is rarely applied in the same way for different organisations, as every business structure varies by its core functions and processes.

Every modelling process uses different steps, algorithms and/or theories to develop a model that can optimise business functions. As a result, FoodBev SETA collaborated with the University of Johannesburg (UJ) to interrogate the 'as-is' processes of the FoodBev industry and provide mechanisms for enhanced efficiency through the design of the 'to-be' processes and a comprehensive systems integration analysis. The aim was to design a comprehensive enterprise architecture that could be applied within the FoodBev enterprises across South Africa and serve as an example of global best practices.

## Digital systems analysis: the University of Johannesburg and the FoodBev SETA project

In the process of achieving the FoodBev SETA's digital optimisation goal, UJ's research team took a software development life cycle (SDLC) approach. This approach includes business landscaping, international best practice reviews and the to-be architecture for SETA. The SDLC is a process of building and maintaining software systems. This approach includes many stages, from preliminary development analysis to post-development software testing and evaluation (Figure 10.1). The UJ team took the following steps.

- First, the UJ team conducted a literature review of global best practices. This exercise informed the approach and the detailed components of this project.

- The UJ team proceeded to conduct a business process analysis through an evaluation of the as-is and to-be processes of the service sector.
- A systems maturity analysis was then conducted to help provide solutions for integrated business processes. This was done through the design of a theoretical architecture for integrated business processes (Figure 10.1).

In terms of mapping the entire project, the UJ team followed the process below.

The UJ team conducted a literature review in the context of the service environment. This was done to gain insights into emerging issues and existing gaps within the literature of service enterprises. With the help of the literature review, the UJ team identified the best practices and potential adoption strategies to develop a to-be architecture. After, a detailed systems analysis and maturity review of the current as-is system's capacity and designs and a critique of relative best practices and opportunities were done. The UJ team also used enterprise business architecture (EBA), a methodology to model, structure and understand business knowledge that assists business managers in making informed decisions. The data collection for the project was done four times a year.

The results of this exercise were twofold. Firstly, an as-is process maturity of all functions of the business in the FoodBev sector was done. The UJ team then conducted a to-be process modelling. In this phase, the potential of the to-be model in such aspects as saving time and resources was assessed. In the second phase, the team reviewed the maturity of the current system. The critique in this phase is based on best practices and to-be processes. The UJ team combined requirements as per the to-be processes and designed an ideal architecture for defining the digital strategy.

## FoodBev as-is maturity and the to-be modelling

The as-is maturity (current processes) was conducted. The existing as-is processes are subject to various analyses resulting in the identification of value-adding prospects, opportunities for creating synergies through redundancy obliteration and optimised processes. 43 core and noncore processes were identified and mapped. Of the 43 processes that were mapped (see Table 10.1), the respective FoodBev teams reviewed all.

Table 10.1 Process maps per department

| Level 1 Category | Process Maps Available | Process Maps |
|---|---|---|
| ET/QA | 5 | 100 |
| Planning | 7 | 100 |
| Research | 11 | 100 |
| Partnership | 20 | 100 |
| TOTAL | 43 | 100 |

## A review of the maturity of the current system

The UJ team's system analyses were based on as-is processes. The team reviewed the as-is processes against the best practices and the to-be practices. Based on the analysis conducted, several proposals for key considerations were made (see Table 10.2), which included the following.

- The ability to host all data in an integrated single enterprise data layer. This would be role-based and all reports from the CEO to all operational units would be available.
- Standard (hourly, daily, weekly, monthly, quarterly and annual reports) should be always available. This implies a central database and BI tool for reports.
- The internal reports must be auto populated with the ability to verify and update.
- External reporting must be automated, with verification. Integration with the Department of Higher Education and Training (DHET) should automate reporting. All DHET data must be available by way of DHET integration. Integration into Sector Education and Training Management Information System (SETMIS) is recommended.
- Further to the above, the team made recommendations on critical areas in the FoodBev sector that can be optimised together with the allocated time per activity.

Table 10.2 Critical areas to consider (continued on next pages)

| | Senior Manager hours | Manager hours | Technical I hours | Intern/Admin hours | Total |
|---|---|---|---|---|---|
| APP | 67 | 66 | 105 | 0 | 237 |
| Develop quarterly plan | 74 | 52 | 64 | 32 | 222 |
| Develop quarterly plan report | 0 | 0 | 0 | 0 | 0 |
| Develop research strategy | 6 | 12 | 20 | 0 | 38 |
| External engagement | 210 | 422 | 467 | 91 | 1189 |
| Internal engagement | 659 | 612 | 0 | 0 | 1271 |
| Monthly report | 150 | 126 | 216 | 48 | 540 |
| Optimise research strategy | 7 | 9 | 12 | 0 | 28 |
| Research agenda Methodology | 12 | 92 | 800 | 84 | 988 |
| Research agenda – Results analysis | 24 | 66 | 408 | 0 | 498 |
| Research instrument | 45 | 75 | 1053 | 156 | 1329 |
| Review research strategy | 6 | 9 | 16 | 0 | 31 |
| SSP Phase 1 | 52 | 82 | 536 | 86 | 756 |
| SSP Phase 2 | 54 | 60 | 295 | 53 | 461 |

Table 10.2 Continued

| | Senior Manager hours | Manager hours | Technical I hours | Intern/Admin hours | |
|---|---|---|---|---|---|
| Statistical analysis | 0 | 8 | 560 | 76 | 644 |
| Strategic plan | 83 | 89 | 138 | 0 | 310 |
| WSP/ATR analysis | 48 | 28 | 150 | 41 | 267 |
| Grand Total | 1496 | 1806 | 4839 | 667 | 8808 |

Considering the vital role of service delivery on the FoodBev SETA's ability to deliver its vision, any fruitful efforts in optimising the delivery model will generate quantifiable advances in outcomes critical to the business imperative. The singular focus on dispensing its directive has permitted limited opportunities to review and optimise FoodBev SETA's delivery model. The complexity of SETA's organisational tasks, processes and expected outcomes arising from the multiplicity of relationships, resources and concomitant interdependencies necessitates persistent improvements to achieve optimal outputs. A comprehensive enterprise architecture that aligns all the business systems is considered beneficial for simplifying the stated complexities. The UJ team proposed comprehensive architecture for the FoodBev.

## The development of a comprehensive architecture

The team proposed that FoodBev begin/start with a high-level architecture from a functional perspective. The high-level architecture design is presented in Figure 10.2 below.

The team added that a few fundamental aspects should be factored in when defining the core systems of the service sector landscape. Table 10.3 shows the proposed aspects in this regard:

Table 10.3 Fundamental aspects for defining the core systems

| | | |
|---|---|---|
| Systems | Supply Chain<br>CRM<br>HR<br>Finance<br>Document Manager | |
| Tools | Workflow engine | |
| Database | Single centralised Database | |
| Portal | External communication/submission | Document |
| Reporting or Business intelligence | Single reporting tool, developed of a central database | |
| External app for student management | | |
| Business process Manager | | |

## Conclusion

Overall, the results and models brought forward by the UJ research team provided valuable insights for scenario planning and informed decision-making for FoodBev. Incorporating the research findings into SETA's strategic planning can help drive digital transformation and position SETA for future success. In conclusion, it is important to integrate systems analysis with business processes for optimising operations and achieving digital enablement. The outcome of this integrated approach and its potential for long-term sustainability and growth for the FoodBev SETA holds much-needed value.

# Exhibits

Figure 10.1 Steps taken by the University of Johannesburg research team (author's own)

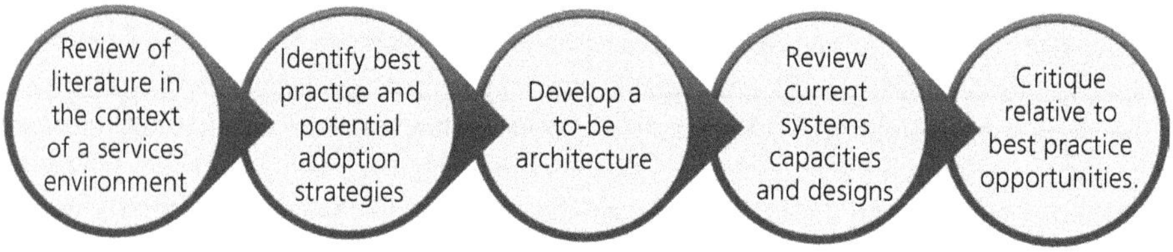


Figure 10.2 High-level architecture (author's own)

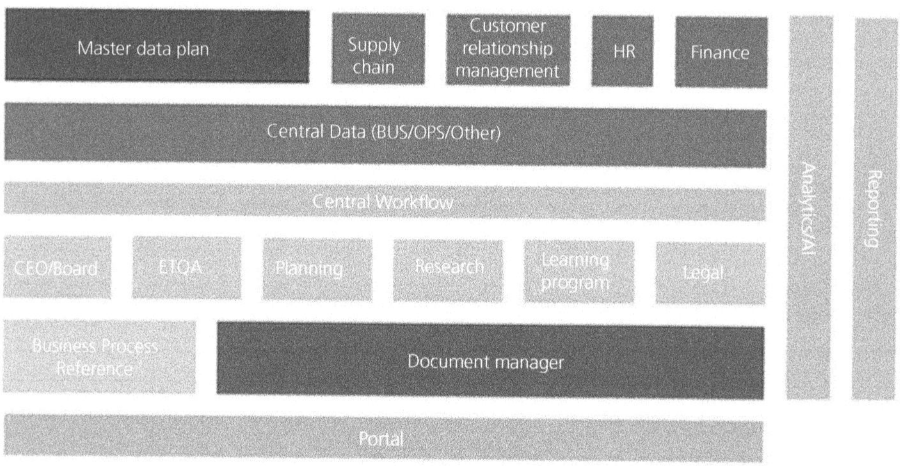


# Teaching notes

## Synopsis

Digitalisation is currently dubbed the key concept in today's business functionality. Within the service sector, the introduction and further development of digital technologies resulted in the transformation of the mode of integration between consumers and producers and different types of organisations. Today, across the globe, the service industry is increasingly experiencing technological novelties. These novelties are informed by both organisational restructurings of existing technologies and the implementation of new technological innovations. In South Africa, the FoodBev SETA acknowledges the digitalisation of the country's FoodBev industry, which is one of the most performing service industries in the country. SETA seeks to optimise and digitalise its operations by way of the adoption of 4IR approaches, including maturity analysis techniques, 4IR strategic approaches and various other techniques. The ability to comprehensively enable best practices is not well documented.

## Teaching objectives

The business management and entrepreneurship modules at both undergraduate and postgraduate levels will find this case study useful. Given that digital technologies and digitalisation are currently the focus of discussions and developments around business growth, competitiveness and sustainability, this case study fulfils the purpose of providing insights on digital systems and maturity analysis to students in their respective relevant disciplines. As such, it is anticipated that, upon completion of reading this case study, the students will be exposed to the following.

- Expertise in digital technologies and the role they play within the service sector's efficiency and sustainability.
- Knowledge of the 'as-is' and 'to-be' pedagogies associated with the digital maturity of a service sector.
- Objective thinking regarding identifying and solving problems associated with digital systems and technologies within the service sector.

## Research methodology

The focus of the case study is to design a digital integrated service for businesses within the FoodBev sector. The case makes use of the software development life cycle (SDLC) approach, which includes the business landscaping, international best practice review and architecture of the 'to-be' architecture for the FoodBev sector. A three-step process was followed within the methodology of this project.

- The first is a literature study or global best practice. This informs the approach and the detailed components of this case study.
- Secondly, a business process analysis was conducted through an evaluation of the 'as-is' and to-be processes of the service sector.
- Thirdly, a system maturity analysis was conducted to help provide solutions for integrated business processes. This was done through the design of a theoretical architecture for integrated business processes.

The above approach helped the UJ team to landscape the capacities of the current vendor systems and provide a critique relative to best practices for this case. The team further identified and added new and technologically aligned methods to the opportunities of the overall systems.

In this case study, a literature review was conducted in the context of the service environment. This was done to gain insights into emerging issues and existing gaps within the literature of service enterprises. With the help of the literature review, the case was able to identify the best practices and potential adoption strategies to develop a to-be architecture. After, a detailed systems analysis and maturity review of current 'as-is' systems capacity and designs were done, as was a critique relative to best practices and opportunities. Further to the above, the enterprise business architecture (EBA) was used.

The EBA is a methodology to model, structure and understand business knowledge that assists business managers in making informed decisions. EBA is a modelling tool to structure related business elements and integrate them into other architectural elements (information, application, technology and security) in the enterprise. It provides a structured phased approach that allows practitioners to move from the as-is state to the future state of the business improvement lifecycle in sequential steps, as below.

- ***Strategy architecture modelling (business strategy formalisation)*** defines the goals of the organisation to provide customer satisfaction with a competitive market edge and captures how stakeholders (internal or external) participate. Tools and approaches such as the balanced scorecard, business motivation model, and strengths, weaknesses, opportunities and threats (SWOT) analysis are utilised. A select list of tools/approaches like the political, economic, social, technological, legal and environmental factors (PESTLE) analysis, product catalog definition, customer segmentation, targeting and positioning, and financial analysis can be used in this phase.
- ***Enterprise process modelling (process architecture)*** describes the structure of process execution knowledge. Process architecture at the enterprise level helps to develop the business context within which the business value chain, value streams and business processes that are performed to achieve business goals are understood. In this phase, an understanding of the business functions, business organisational structure and the linkage between business processes and other business elements such as business events, organisational role players, business units and business locations is generated.
- ***Alignment of strategy to process*** enables the measurement of performance. This 'architecture alignment' phase helps in monitoring business strategy execution. The elements of the strategy must be deconstructed into operational aspects and then tracked for effective performance. Performance measures need to be defined and associated with business goals.
- ***Business architecture analysis*** provides the feedback necessary to perform ongoing improvements to the strategic architecture. The multiple business scenarios that an enterprise faces can be broadly grouped into business reorganisation scenarios, business improvement scenarios, business change scenarios and IT-led business transformation scenarios.
- ***Maturity analysis*** enables insights into the current state of digital implementation. This helps to formulate relevant solutions for digital optimisation.
- ***Developing an integrated digital strategy*** helps in providing guidelines and addressing existing gaps in the lack of integrated systems, which is still a big challenge in current business processes.

**Keywords**
Digital Enablement; Service Enterprise; Systems Analysis; FoodBev.

## Assignment questions

- Evaluate the economic contributions of the FoodBev sector.
- What are the key challenges faced due to the limited ability to comprehensively enable best practices within the food sector?
- Examine the significance of implementing comprehensive business processes and maturity analysis within the FoodBev sector.
- Provide a distinction between the 'as-is' and 'to-be' processes and articulate why this is crucial for a service enterprise.

## Teaching plan and timing

For a class of 90 minutes, the researcher recommended the following teaching plan.

- *5 minutes*: Introduction. The instructor should provide a summary of the case.
- *80 minutes*: Teaching of the assignment questions. This should be structured in the following manner:
- *20 minutes:* Evaluate the economic contributions of the FoodBev sector.
- *20 minutes:* What are the key challenges faced due to the limited ability to comprehensively enable best practices within the FoodBev sector?
- *20 minutes:* Examine the significance of implementing a comprehensive business process and maturity analysis within FoodBev.
- *20 minutes:* Provide a distinction between the 'as-is' and 'to-be' processes and articulate why this is crucial for a service enterprise.
- *5 Minutes:* Conclusion. Wrap up the case and summarise key learnings.

## Analysis of assignment questions

**1. Evaluate the economic contributions of the FoodBev sector.**

The value proposition of this question is to teach learners about the significance of the FoodBev sector to the economy.

*[Teaching suggestion]* The instructor can ask the learners to identify different economic contribution factors of the FoodBev sector. Some of the factors include the following:

The FoodBev sector contributes to employment creation in many destinations. The sector is one of the biggest employers, responsible for employing people with low- to high-level skills. Given the high rates of unemployment, especially in developing countries, the FoodBev sector, with its ability to absorb even the less skilled people, assists in reducing those statistics.

The FoodBev sector can contribute to the multiplier effect of a destination. The most utilised service is FoodBev, given that despite the differences in activities that people engage in any given area, they will still need to eat and/or drink. Because of that, there is more spending on FoodBev, resulting in an increased multiplier for the sector. The sector, therefore, plays a major role in contributing to economic growth and sustainability.

**2. What are the key challenges faced due to the limited ability to comprehensively enable best practices within the FoodBev sector?**

The instructor should provide insights to learners on the key challenges that are likely to be experienced due to the lack of ability to comprehensively enable best practices within the FoodBev sector.

*[Teaching suggestion]* The instructor can ask the learners to identify associated challenges that may arise from a limited ability to comprehensively enable best practices within the FoodBev sector. Before getting into the discussion, it is crucial to enlighten the learners on what best practices mean. A best practice is a method or technique that has been generally accepted as superior to other known alternatives because it often produces results that are superior to those achieved by other means or because it has become a standard way of doing things.

In today's business environment, digital technologies and digital maturity dictate the pace of current best practices in conducting business. In the FoodBev sector, best practices are the ones that promote increased efficiency, growth of the sector, and, eventually, a meaningful contribution to economic development and growth.

Slowed growth and transformation of the FoodBev sector: as the world continues to evolve, new technologies are invented to ensure there is a positive outcome in the production of goods and services. Only those businesses that are quick to adapt to these changes will see much-needed growth. Above that, consumers are also attracted to purchasing and consuming goods and products offered more conveniently. Businesses that fail to make it easier for consumers to access their offerings tend to fail. Therefore, a limit on enabling new technologies may result in the FoodBev businesses' slow growth and transformation, as they will be considered unattractive and lack the skill to adapt to environmental changes.

The above situation will result in the FoodBev sector not being able to fulfil its role in economic growth and transformation, especially in developing economies. Also, this will affect the role of the sector in the overall economic sustainability of a destination,

**3. Examine the significance of implementing comprehensive business processes and maturity analysis within the FoodBev sector.**

The value proposition of this question is to educate the learners on how to examine the importance of implementing comprehensive business processes and maturity analysis within FoodBev. Also, it is crucial to teach the dangers of not doing so.

*[Teaching Suggestion]* Comprehensive business processes and maturity analysis can also be viewed as strategic transformations targeting organisational changes implemented through digitalisation projects to enable major business improvements. Many organisations have recently begun to do a maturity analysis of their systems and are now implementing solutions based on digital technologies. It is crucial to redesign business functions incorporating digital technologies and to ensure that digital processes are put in place to increase business

efficiency. This allows businesses to be future-ready and be able to increase profits and growth more than businesses that do not prioritise business maturity.

A maturity analysis of digital optimisation in businesses implies that there is a business transformation that is focused on digital capabilities to transform a business into a competitive and sustainable entity within the digital economy. Recent trends in businesses' growth strategies are that many businesses across the globe are prioritising digital technologies to efficiently strategise their business functions and increase competitiveness.

**4. Provide a distinction between the 'as-is' and 'to-be' processes and articulate why this is crucial for a service enterprise.**

*[Teaching suggestion]* The instructor should probe learners to discuss the differences between the 'as-is' and the 'to-be' processes. The 'as-is' processes include the current business processes, that are, in most cases, less efficient in moving the business towards a sustainable direction. On the other hand, 'to-be' processes include the proposed best practices that may be able to assist in the organisation's efficiency. The key differences between the as-is and to-be processes are summarised in Table 10.4.

Table 10.4 Key differences between the as-is and to-be processes

| As-is processes | To-be processes |
|---|---|
| A detailed overview of the current state of a company's process, culture and capabilities. | Outlines how a company's process, culture and capabilities will appear in the future. |
| Outlines how work is currently performed and how information flows through the organization. | It serves as a roadmap for change, identifying where the organisation needs to be to achieve its goals. |
| The purpose of an as-is business process is to provide a baseline for identifying areas for improvement, optimising the use of resources and ultimately increasing efficiency and productivity. | The purpose of a to-be business process is to provide a clear picture of what the future state of the organisation will look like and what steps need to be taken to achieve it. |

## Key learnings

The instructor can conclude the teaching by highlighting the importance of the digital enablement of the service sector. It is also crucial to provide a pedagogy on the significance of a comprehensive systems and maturity analysis and the associated benefits of doing so for the service sector – for example, increased proficiency and business functionality, increased competition, and, ultimately, economic growth and transformation are some of the key outcomes for the digital enablement of a service sector.

REFERENCES

Ghosh A, Nashaat M and Miller J (2019) The current state of software license renewals in the I.T. industry. *Information and Software Technology* **108**: 139–152.

Arnesh Telukdarie
ISBN 978-1-83608-365-8
https://doi.org/10.1108/978-1-83608-362-720241013


# Chapter 11
# Final reflections on the journey of integrated digital transformation in engineering management

A Telukdarie and H Mahure

This book dived into the fascinating fields of engineering management and digital transformation. In today's digital age, achieving a digitally integrated business can be challenging and often requires practical guidance for professionals to gain confidence as they embark on their digital transformation journey. To bridge this gap, this book has offered diverse and detailed perspectives through real-world case studies, documenting the journey of industry professionals on their digital transformation journey. The case studies offered a comprehensive and detailed account of the path to effective digital transformation, highlighting the challenges and successful strategies employed by industry leaders. These firsthand accounts offered practical experiences and learnings from the case studies to assist in future digital implementations.

The book proves to be an invaluable resource for readers seeking to learn from the real-world experiences of industry professionals, providing them with the insights required to make informed decisions and drive successful digital transformation in their organisations. By reading this book, professionals can gain a deeper understanding of the complexities involved in digital transformation and how to navigate them successfully.

This book consists of nine chapters that delve into the various aspects of integrated digital transformation for different sectors. Figure 11.1 displays a summary of the industries that the case studies covered in this book.

Each chapter provides sector-specific knowledge, covering diverse and distinct aspects of digital transformation. The first chapter focused on the implementation of a business intelligence system at the Chemical Industries Education and Training Authority in South Africa. This chapter offered a comprehensive overview of how business intelligence can enhance skills in forecasting, data analytics, and decision-making processes. It also highlighted the significance of data-driven strategies for aligning training programs with labour market requirements and achieving long-term success in skills development initiatives.

The second chapter delved into the topic of digital transformation in the health care sector, providing practical guidelines on how integrated digital innovation can lead to cost-effective

Figure 11.1 Digital transformation in sectors (author's own)

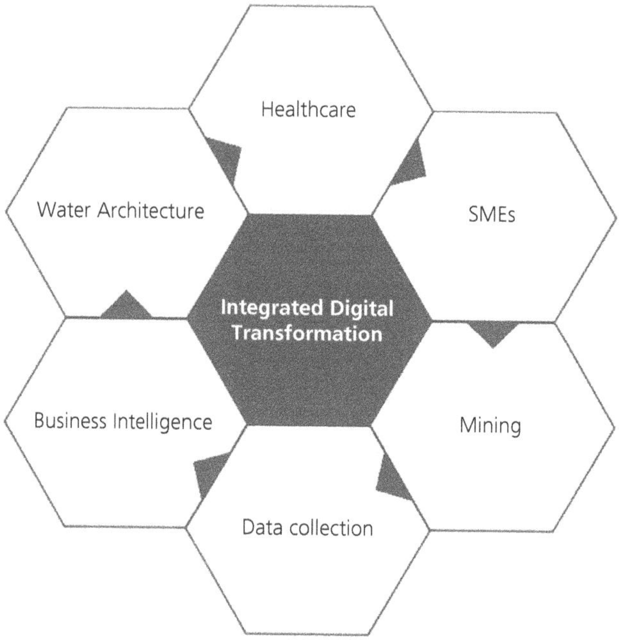


optimisation. The chapter outlined the steps taken to develop the digital solution and demonstrated how innovative methods and digital tools can be used to solve pressing business challenges systematically. The chapter delves deeper into the health care sector, exploring the various challenges faced by health care providers and how digital transformation can revolutionise the industry. The chapter provided insights into how digital transformation can help health care providers improve patient outcomes, reduce costs, and enhance the overall quality of health care services.

Chapter three of the book provided practical guidelines for integrated digital transformation in the BBT steel industry. The chapter explored BBT steel manufacturers and the challenges they face in forecasting due to a gap between their manufacturing and business systems. By utilising existing investments to create smart enterprises, integrated digitisation played a crucial role in bridging this gap, enabling efficient management of current and future steel. The chapter proposed a framework that can be implemented in the steel sector and provided guidelines on how to adopt Industry 4.0 integrated technologies.

Chapter four shed light on the challenges and possibilities faced by small and medium enterprises (SMEs) in South Africa's food and beverage manufacturing industry during their shift towards digitalisation. The chapter provided guidelines on how SMEs can overcome industry challenges, including competition with larger businesses, financial limitations, limited access to technology, credit difficulties, skill shortages, and policy restrictions.

Chapter five of the book explored the topic of digitalisation and its role in collecting data in the context of childhood development in Africa. The chapter provided guidelines on how industry professionals and researchers can move away from the traditional paper-based data collection system to a sophisticated digital tool for data collection and analysis. The case study highlighted the challenges and opportunities involved in transitioning to a digital system, including the advantages of greater efficiency, accuracy, and speed of data processing.

In Chapter six, the book analysed the challenges Company X faced during the digital transformation process. The chapter documented the development of a comprehensive digital business strategy aimed at achieving total business visibility, automation of non-value-adding activities, and digitalisation of various aspects. The case stated in detail the benefits of digitalisation and highlighted the need for an integrated approach to digital transformation.

Chapter seven of the book provided guidelines on an integrated digitalisation approach in water architecture. It explored how Johannesburg, a city grappling with water scarcity and distribution challenges, leverages advanced fourth industrial revolution technologies to overcome these hurdles.

Moving on to Chapter eight, the book analysed the implementation of a cyber-physical system (CPS) in a multinational energy company. This chapter provided an in-depth understanding of how CPS was adopted, and the benefits derived from its implementation.

Chapter nine of the book explored the food and beverages architecture in South Africa. It provided a detailed account of how 4IR technologies enhance the efficiency, and effectiveness, of the FoodBev Sector Education and Training Authority.

Overall, this book has proven to be an excellent resource for anyone interested in learning about the practical implementation of an integrated digital business across various industries. It has offered valuable insights into how digitalisation can enhance business efficiency and sustainability. The book provides an in-depth and diverse perspective on digital transformation through real-world case studies of industry professionals who navigate the complexities of digitalisation. These case studies give valuable insights into the effective path of integrated digital transformation, highlighting the challenges, pitfalls, and successes of real industry professionals.

This book is a valuable tool to assist future professionals in their digitalisation journey by sharing the lessons learned from practical experience. It has also contributed towards building Africa's business environment and development through practical examples of local solutions to a range of business challenges. 'Exploring Digitalisation in Engineering Management' is an essential guide for both practitioners and researchers who want to understand the transformative potential of digitalisation.

Arnesh Telukdarie
ISBN 978-1-83608-365-8
https://doi.org/10.1108/978-1-83608-362-720241014


# Index

www.ingramcontent.com/pod-product-compliance
Ingram Content Group UK Ltd.
Pitfield, Milton Keynes, MK11 3LW, UK
UKHW020047060325
455911UK00004B/51

9 781836 083658